ELECTRONICS POCKET HANDBOOK

Daniel L. Metzger

Monroe County Community College

PRENTICE-HALL, INC.,
ENGLEWOOD CLIFFS, N.J. 07632

*Library of Congress Cataloging
in Publication Data*

Metzger, Daniel L.,
    Electronics pocket handbook.

    Includes index.
    1. Electronics—Handbooks, manuals, etc. I. Title.
TK7825.M45   621.381'0212   81-15410
ISBN 0-13-251835-X   AACR2

Editorial/production supervision
and interior design by *Mary Carnis*
Cover design by *Edsal Enterprises*
Page layout by *Cathy Colucci*
Manufacturing buyer: *Gordon Osbourne*

Printed in the United States of America

10 9 8 7 6 5 4

ISBN   0-13-251835-X

Prentice-Hall International, Inc., *London*
Prentice-Hall of Australia Pty. Limited, *Sydney*
Prentice-Hall of Canada, Ltd., *Toronto*
Prentice-Hall of India Private Limited, *New Delhi*
Prentice-Hall of Southeast Asia Pte. Ltd., *Singapore*
Whitehall Books Limited, *Wellington, New Zealand*

Contents

ELECTRONICS
POCKET
HANDBOOK

1

Definitions, Formulas, and Charts

1.1 DEFINITIONS OF ELECTRICAL UNITS

Ampere (A). The current (I) that would produce a force of 2×10^{-7} newton per meter of length between two parallel conductors 1 meter apart. The fundamental electrical unit in the metric (SI) system. Equal to a charge flow rate of 1 coulomb per second.

Coulomb (C). The charge (Q) that passes by a point in 1 second when the current is 1 ampere. Equals the charge of 6.242×10^{18} electrons.

Joule (J). The work (W) done by a force of 1 newton acting through a distance of 1 meter.

Watt (W). The power (P) required to do work at a rate of 1 joule per second.

Volt (V). The potential difference between two points on a wire carrying 1 ampere of current when the power dissipated between the points is 1 watt.

Ohm (Ω). The resistance (R) that produces a voltage of 1 volt when carrying a current of 1 ampere.

Siemens (S). The conductance (G) that produces a current of 1 ampere when the applied voltage is 1 volt. Reciprocal ohm. Formerly *mho* ($\mho$).

Farad (F). The capacitance (C) in which a charge of 1 coulomb produces a potential difference of 1 volt.

Henry (H). The inductance (L) in which 1 volt is induced for a current rate of change of 1 ampere per second.

1.2 OHM'S LAW AND POWER LAW FOR RESISTIVE ELEMENTS

The commonly measured electrical quantities are voltage, current, resistance, and power. If any two of these are known, the other two can be calculated.

$$I = \frac{V}{R} = \frac{P}{V} = \sqrt{\frac{P}{R}}$$

$$V = IR = \frac{P}{I} = \sqrt{PR}$$

$$R = \frac{V}{I} = \frac{P}{I^2} = \frac{V^2}{P}$$

$$P = IV = I^2R = \frac{V^2}{R}$$

1.3 RESISTOR COMBINATIONS

Series resistors may be replaced by a single resistor equal to their sum. Resistors are in series if the identical current flows in each of them, regardless of whether they are actually connected end to end.

$$R_T = R_1 + R_2 + R_3 + \cdots$$

Parallel resistances may be converted to conductance units and reduced to a single resistance having a conductance equal to the sum of the conductances. Resistors are in parallel if the identical voltage appears across them, regardless of whether they are actually connected across one another.

$$G_T = G_1 + G_2 + G_3 + \cdots$$

$$R_T = \frac{1}{\dfrac{1}{R_1} + \dfrac{1}{R_2} + \dfrac{1}{R_3} + \cdots}$$

With a calculator, proceed as follows:

$$R_1 \boxed{1/x} \boxed{+} \quad R_2 \boxed{1/x} \boxed{+} \quad R_3 \boxed{1/x} \boxed{=} \boxed{1/x}$$

Parallel design: To find R_2 such that R_2 in parallel with R_1 will produce a desired R_T :

$$R_2 = \frac{1}{\dfrac{1}{R_T} - \dfrac{1}{R_1}}$$

1.4 INDUCTORS AND CAPACITORS

Inductor. An element that reacts against any change in current through it by generating a voltage in opposition to the applied voltage and proportional to the rate of change of current.

Capacitor. An element that opposes any change in voltage across it by passing a current proportional to the rate of change in voltage.

Table 1-1 lists the major formulas for inductors and capacitors.

3

TABLE 1-1. Inductor and Capacitor Formulas

Capacitors

Charge relationship	$Q = CV$
Reaction against changing V	$I = C\dfrac{\Delta V}{\Delta t}$
Constant charge and discharge	$CV = It$
Energy stored	$W = \frac{1}{2}CV^2$
Time constant	$\tau = RC$
Series combination	$C_T = \dfrac{1}{\dfrac{1}{C_1} + \dfrac{1}{C_2} + \cdots}$
Parallel combination	$C_T = C_1 + C_2 + \cdots$

Inductors

Charge relationship	$Q = It$
Reaction against changing I	$V = L\dfrac{\Delta I}{\Delta t}$
Constant V charge and discharge	$LI = Vt$
Energy stored	$W = \frac{1}{2}LI^2$
Time constant	$\tau = \dfrac{L}{R}$
Series combination (no mutual inductance)	$L_T = L_1 + L_2 + \cdots$
Parallel combination	$L_T = \dfrac{1}{\dfrac{1}{L_1} + \dfrac{1}{L_2} + \cdots}$

Mutual inductance is possessed by two coils whose magnetic fields are coupled. Measure L_{AID} with the two coils in series aiding, and L_{OPP} with the two coils in series opposing.

$$M = \frac{L_{\text{AID}} - L_{\text{OPP}}}{4}$$

Where M is known, two inductors in series have the total

$$L_T = L_1 + L_2 + 2M \text{ (aiding fields)}$$
$$L_T = L_1 + L_2 - 2M \text{ (opposing fields)}$$

The coefficient of coupling of two coils is

$$k = \frac{M}{\sqrt{L_1 L_2}}$$

Time constant (τ) is the time required for a process to go 63.2% of the way to completion following the rising curve, given by

$$y = 1 - e^{-t/\tau}$$

or the falling curve

$$y = e^{-t/\tau}$$

where y represents the fraction of maximum voltage or current. Figure 1-1 presents these curves graphically. Notice that the initial charge or discharge rate would, if continued, bring the process to completion in one τ. Figure 1-2 shows how the basic time-constant curve applies to RL and RC circuits.

1.5 AC FORMULAS

Radian frequency:

$$\omega = 2\pi f \qquad f = \frac{\omega}{2\pi}$$

Period and frequency:

$$T = \frac{1}{f} \qquad f = \frac{1}{T}$$

Rising curve

$\dfrac{t}{\tau}$	y
0.1	0.095
0.2	0.181
0.5	0.393
1.0	0.632
2.0	0.865
3.0	0.950
4.0	0.982
5.0	0.993
7.0	0.999

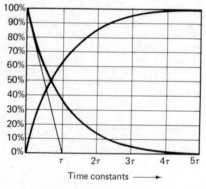

Time constants ⟶

FIGURE 1-1

Inductor charging

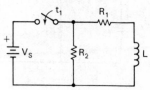

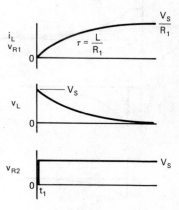

FIGURE 1–2

Inductor discharging

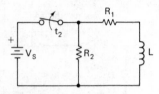

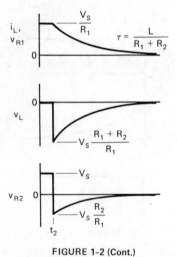

FIGURE 1-2 (Cont.)

8

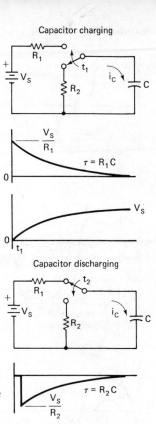

Capacitor charging

i_C, v_{R1}

$$\frac{V_S}{R_1}$$

$\tau = R_1 C$

0

v_C

V_S

0

t_1

Capacitor discharging

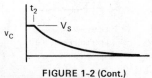

i_C, v_{R2}

$$\frac{V_S}{R_2}$$

$\tau = R_2 C$

v_C

t_2

V_S

FIGURE 1-2 (Cont.)

Harmonic distortion (see Fig. 1-3)

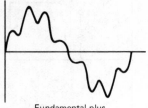

Fundamental plus
7th harmonic

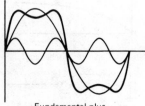

Fundamental plus
3rd harmonic

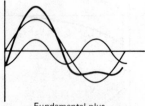

Fundamental plus
2nd harmonic

FIGURE 1-3

Fourier series of common waveforms (see Fig. 1-4)

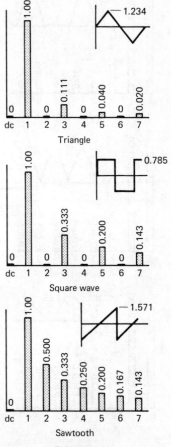

FIGURE 1-4

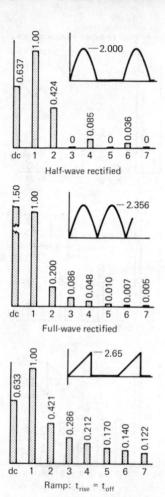

Half-wave rectified

Full-wave rectified

Ramp: $t_{rise} = t_{off}$

FIGURE 1-4 (Cont.)

12

Peak, average, and rms:

1. Use peak voltage or current to calculate maximum *instantaneous* power only. Use V^2/R or I^2R.

2. Use average current to calculate average power when the voltage is fixed dc. Use average voltage to calculate average power when the current is unvarying dc.

3. Use rms voltage and/or rms current to calculate average power when the load is a linear device and both V and I are ac in phase and of the same waveshape. Use IV, I^2R, or V^2/R.

4. Rms measure is assumed in any ac voltage or current notation unless peak, peak to peak, or average is specified.

5. The factor 0.707 for converting peak to rms applies to sine waves only. Figure 1-5 gives the factors for selected waveshapes.

6. VOMs and most DVMs are calibrated to read $1.11\ V_{\text{avg}}$ of an ac wave. VTVMs and FET meters generally read $0.707V_{\text{pk}}$. For nonsinusoidal waveforms the resultant reading is *not* V_{rms}.

The rms value of a nonsinusoidal wave V_t can be obtained from the rms values of its harmonic components V_n:

$$V_t = \sqrt{V_1{}^2 + V_2{}^2 + \cdots + V_n{}^2}$$

Skin effect (Fig. 1-6) seriously increases the resistance of a copper wire in free space when the skin depth δ becomes less than one-half the diameter of the wire:

$$\delta = \frac{66}{\sqrt{f}}$$

where δ is in millimeters and f is in hertz. Wires in bundles or wound as coils suffer from skin effect at substantially lower frequencies because of *proximity effect*.

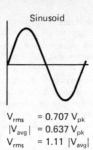

Sinusoid

$$V_{rms} = 0.707 \ V_{pk}$$
$$|V_{avg}| = 0.637 \ V_{pk}$$
$$V_{rms} = 1.11 \ |V_{avg}|$$

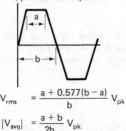

Symmetrical trapezoid

$$V_{rms} = \frac{a + 0.577(b - a)}{b} \ V_{pk}$$
$$|V_{avg}| = \frac{a + b}{2b} \ V_{pk}$$

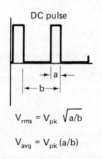

DC pulse

$$V_{rms} = V_{pk} \ \sqrt{a/b}$$
$$V_{avg} = V_{pk} \ (a/b)$$

FIGURE 1–5

14

Triangle or sawtooth

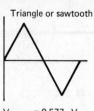

$V_{rms} = 0.577 \; V_{pk}$
$|V_{avg}| = 0.500 \; V_{pk}$
$V_{rms} = 1.154 \; |V_{avg}|$

Sine wave on dc level

$$V_{rms} = \sqrt{V_{DC}^2 + \frac{V_{pk}^2}{2}}$$

Square wave

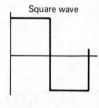

$V_{rms} = V_{pk}$
$|V_{avg}| = V_{pk}$

FIGURE 1–5 (Cont.)

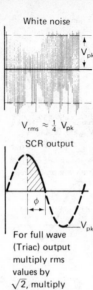

White noise

$V_{rms} \approx \frac{1}{4} V_{pk}$

SCR output

ϕ

V_{pk}

For full wave
(Triac) output
multiply rms
values by
$\sqrt{2}$, multiply
avg values by 2

ϕ	V_{rms}/V_{pk}	V_{avg}/V_{pk}
15°	0.028	0.008
30°	0.085	0.022
45°	0.15	0.045
60°	0.22	0.078
75°	0.29	0.115
90°	0.35	0.16
105°	0.41	0.21
120°	0.45	0.24
135°	0.48	0.27
150°	0.49	0.29
165°	0.50	0.31
180°	0.50	0.32

FIGURE 1-5 (Cont.)

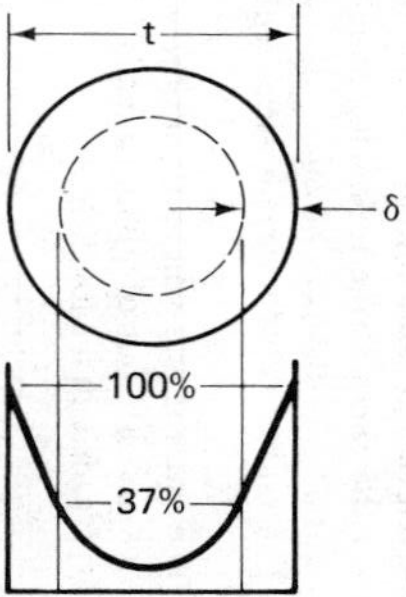

Skin effect increases R less than 10%
for wires smaller than AWG No:

f	Free space	Single layer	Multilayer
60 Hz	–	0	8
400 Hz	00	4	11
1 kHz	3	6	14
10 kHz	13	14	27
100 kHz	22	32	44
1 MHz	32	44	–
10 MHz	44	–	–

FIGURE 1-6

The resistance of a copper wire in free space under the skin effect is

$$R = 8.3 \times 10^{-5} \, d \, \frac{\sqrt{f}}{t}$$

where d is length in meters, t is diameter in millimeters, and f is in hertz. For other metals, multiply R by the square root of the resistivity relative to copper. The resistance given by the foregoing equation should be multiplied by approximately 10 for single-layer close-wound coils, and by 50 to 100 for multilayer coils.

1.6 REACTIVE AC FORMULAS

Reactance (X):

1. The concept of reactance applies to sinusoidal waveforms only. Other waveforms must be treated as a fundamental plus harmonic sine waves. Reactive networks will affect each of these components differently, generally resulting in distortion of the waveshape.
2. Reactive devices store energy on the first quarter-cycle of ac and reflect it back to the source on the second quarter-cycle. They limit current but do not dissipate power.
3. The reactance of a device varies with frequency.
4. The current and voltage in a purely reactive device are 90° out of phase: v leads i in an inductor, i leads v in a capacitor.

Reactance of inductor and capacitor:

$$X_L = 2\pi f L \qquad X_C = \frac{-1}{2\pi f C}$$

Ohm's law for reactance:

$$I = \frac{V}{X} \qquad X = \frac{V}{I} \qquad V = IX$$

Susceptance is negative reciprocal reactance:

$$B = \frac{-1}{X}$$

Reactances in series add algebraically:

$$X_T = X_1 + X_2 + X_3 + \cdots$$

Note that capacitive reactances *subtract* from inductive reactances.

Reactances in parallel: change to susceptances and add:

$$X_T = \frac{1}{\dfrac{1}{X_1} + \dfrac{1}{X_2} + \dfrac{1}{X_3} + \cdots}$$

Resonance (tuned circuit): Series LC combination has zero reactance at resonant frequency f_r. Parallel LC has infinite reactance at f_r.

$$f_r = \frac{1}{2\pi\sqrt{LC}}$$

A reactance chart (Fig. 1-7) can be used to estimate:

1. Resonant frequency: $1\mu F$ and 0.5 H resonate near 230 Hz in the example point.
2. Reactance vs. frequency: $1\mu F$ (or 0.5 H) has a reactance of about 700Ω at 230 Hz in the example point.

Impedance (Z) is the combination of resistance and reactance. The phase angle θ between voltage and current is between 0 and $\pm90°$ and the power dissipated is between IV and 0.

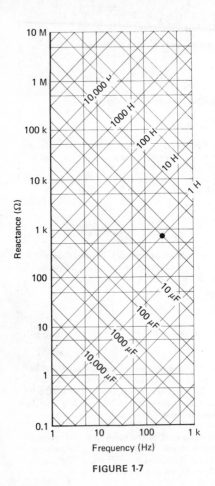

FIGURE 1-7

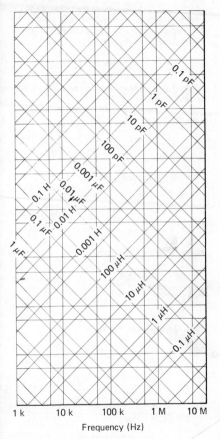

FIGURE 1-7 (Cont.)

Power factor (F_p) is the ratio of true power P (given by I^2R) dissipated in the resistance to the "apparent power" S, which is the IV product for the circuit, neglecting the phase difference between I and V. In general:

$$F_p = \cos\theta = \frac{P}{S}$$

where θ is the phase angle between circuit voltage and current.

Admittance (Y) is reciprocal impedance:

$$Y = \frac{1}{Z}$$

Ohm's law for impedance:

$$I = \frac{V}{Z} \qquad Z = \frac{V}{I} \qquad V = IZ$$

$$P = IV\cos\theta \qquad P = I^2Z\cos\theta \qquad P = \frac{V^2}{Z}\cos\theta$$

Series RX circuit [Fig. 1-8(a)]:

$$Z = \sqrt{X^2 + R^2}$$

$$\theta = \tan^{-1}\frac{X}{R} \qquad F_p = \cos\theta = \frac{R}{Z}$$

Parallel RX circuit [Fig. 1-8(b)]:

$$Z = \frac{1}{G^2 + B^2} = \frac{RX}{X^2 + R^2} \qquad \theta = \tan^{-1}\frac{R}{X}$$

$$F_p = \cos\theta = \frac{Z}{R}$$

Parallel-series RX conversion [Fig. 1-9(a)] (valid only at single frequency):

$$R_s = X_p\,\frac{X_pR_p}{X_p^2 + R_p^2} \qquad X_s = R_p\,\frac{X_pR_p}{X_p^2 + R_p^2}$$

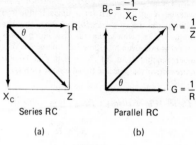

FIGURE 1-8

Series-parallel RX conversion [Fig. 1-9(b)]:

$$R_p = \frac{X_s^2 + R_s^2}{R_s} \quad X_p = \frac{X_s^2 + R_s^2}{X_s}$$

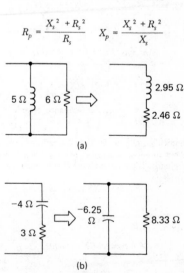

FIGURE 1-9

RLC L-networks can be used to transform a resistance R up or down (at a single frequency) without the use of a transformer. Note that the frequency at which the reactive components cancel is not given by $1/2\pi\sqrt{LC}$ (unless R is very much smaller than X). Note that Fig. 1-10(a) represents a real parallel tuned circuit with $Q_L = X_L/R$.

Design [Fig. 1-10(a)]   *Design [Fig. 1-10(b)]*

$R_{in} > R_L$ $R_{in} < R_L$

$X_L = \sqrt{R_L R_{in} - R_L^2}$ $X_L = \dfrac{R_L R_{in}}{X_C}$

$X_C = \dfrac{R_L R_{in}}{X_L}$ $X_C = R\sqrt{\dfrac{R_{in}}{R_L - R_{in}}}$

Analysis [Fig. 1-10(a) and (b)]

$$Z_{in} = X_C \sqrt{\frac{R^2 + X_L^2}{R^2 + (X_L - X_C)^2}}$$

$$Z_{in} = \frac{X^2}{R} = QX \left(\begin{array}{c} \text{at resonance, error} \\ < 2\% \text{ for } Q > 5 \end{array} \right)$$

Bandwidth, Q, and D: Q is the ratio of energy stored to energy dissipated per cycle in a device or circuit. Dissipation factor D is its reciprocal.

For series RX elements:

$$Q = \frac{X_s}{R_s}$$

For parallel RX elements:

$$Q = \frac{R_p}{X_p}$$

24

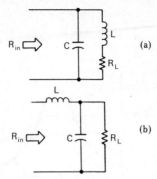

FIGURE 1-10 L-networks

For a series resonant circuit:

$$Q = \frac{X_C}{R_s} = \frac{X_L}{R_s}$$

For a parallel resonant circuit with parallel resistance:

$$Q = \frac{R_p}{X_C} = \frac{X_p}{X_L}$$

For a parallel resonant circuit with series resistance in the inductor, [Fig. 1-10(a)], an approximation to within 4% for $Q_{coil} > 5$ is

$$Q = \frac{X}{R_s}$$

The Q of a tuned circuit provides an estimate of its *bandwidth* (B), which is the frequency span between the upper and lower half-power points, as shown in Fig. 1-11.

$$B = \frac{f_r}{Q}$$

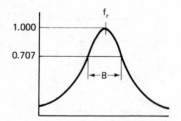

Bandpass: series *LC* in the line
or parallel *LC* across the line

(a)

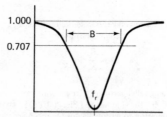

Trap: series *LC* across the line
or parallel *LC* in the line

(b)

FIGURE 1-11

1.7 COMPLEX NOTATION: RECTANGULAR AND POLAR (PHASOR) FORMS

Rectangular Form is $R \pm jX$, where R is resistance in ohms, X is reactance (positive for inductive, negative for capacitive,) and $j = \sqrt{-1}$. R is called the *real* part and X the *imaginary* part. Figure 1-12 gives examples of circuits and their complex expressions.

Add and subtract real and imaginary parts separately:

$$(3 + j6) + (2 - j4) = 5 + j2$$

Multiply and divide both parts by real numbers as usual:

$$3(1 + j2) = 3 + j6$$

$$\frac{4 - j6}{-2} = -2 + j3$$

Multiply or divide by a j term; remember that $j^2 = -1$, $j^3 = -j$, and $j^4 = +1$:

$$-j3(5 + j) = -j15 - j^2 3 = 3 - j15$$

$$\frac{9 - j3}{j3} = \frac{j^4 9 - j3}{j3} = j^3 3 - 1 = -1 - j3$$

Multiply complex terms by conventional algebra; product of *first* terms + product of *outer* terms + product of *inner* terms + product of *last* terms:

$$(8 + j3)(2 - j5) = 16 - j40 - j6 - j^2 15 = 31 - j34$$

Divide complex terms by multiplying both terms by the complex conjugate of the denominator. This converts the denominator to a real number. Form the complex conjugate by changing the sign of the j term.

$$\frac{5 + j10}{1 - j2} \times \frac{1 + j2}{1 + j2} = \frac{5 + j10 + j10 + j^2 20}{1 - j^2 4} =$$

$$\frac{-15 + j20}{5} = -3 + j4$$

Division and multiplication are easier in polar form: Polar form expresses magnitude and phase angle, wheras rectangular form expresses real (resistive or in-phase) and imaginary (reactive or 90° out-of-phase) components (see Fig. 1-8). $R + jX = Z\underline{/\theta}$.

To convert rectangular to polar form:

$$Z = \sqrt{R^2 + X^2}$$
$$\theta = \tan^{-1} \frac{X}{R}$$

To convert polar to rectangular form:

$$R = Z \cos \theta$$
$$X = Z \sin \theta$$

To multiply polar quantities, multiply the magnitudes but add the phase angles algebraically:

$$2 \underline{/30°} \times 4 \underline{/-25°} = 8 \underline{/5°}$$
$$3 \times 5 \underline{/60°} = 15 \underline{/60°}$$

To divide polar quantities, divide the magnitudes but subtract the denominator phase angle from the numerator phase angle algebraically:

$$\frac{12 \underline{/15°}}{4\underline{/-30°}} = 3 \underline{/45°}$$

Using the example from the previous complex division:

$$\frac{5 + j10}{1 - j2} = \frac{11.18 \underline{/63.4°}}{2.24 \underline{/-63.4°}} = 5 \underline{/126.9°} = -3 + j4$$

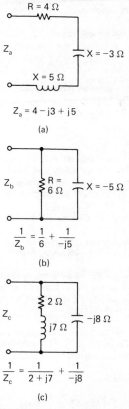

FIGURE 1-12 Complex expressions for three circuits

29

1.8 TRANSFORMER FORMULAS

Turns ratio: $n = N_S/N_P$, where N_S and N_P are secondary and primary number of turns.

Voltage and current ratios:

$$n = \frac{V_S}{V_P} = \frac{I_P}{I_S}$$

Impedance ratio, ideal transformer, unity coupling k (Fig. 1-13):

$$\frac{Z_S}{Z_P} = n^2 = \left(\frac{V_S}{V_P}\right)^2$$

Z_P is the impedance reflected across to the primary when the secondary load is Z_S. Z_S is the source impedance at the secondary when the primary source impedance is Z_P.

Low-frequency cutoff of a transformer:

$$f_{\text{low}} = \frac{r_g R_L}{2\pi L_P (n^2 r_g + R_L)}$$

where f_{low} is the –3-dB point, r_g the generator source resistance, R_L the secondary load resistance, L_P the primary inductance in henries, and n the turns ratio N_S/N_P.

Transformer saturation is observed on an oscilloscope as gross distortion of the secondary waveform from that applied by the source.

Sine-wave-driven:

$$\frac{V_1(\text{max})}{f_1} = \frac{V_2(\text{max})}{f_2}$$

Thus, a transformer that saturates at 15 V at 6 Hz will saturate at 150 V at 60 Hz.

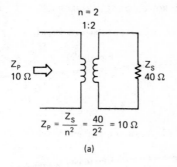

$$Z_P = \frac{Z_S}{n^2} = \frac{40}{2^2} = 10 \ \Omega$$

(a)

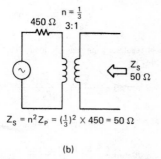

$$Z_S = n^2 Z_P = \left(\frac{1}{3}\right)^2 \times 450 = 50 \ \Omega$$

(b)

FIGURE 1-13 Impedance transformation

31

Pulse-driven: $V_1 t_1 = V_2 t_2$. Thus, a pulse transformer that can take 5 V for $30\mu s$ can take 10 V for $15\mu s$.

1.9 POWER-SUPPLY FORMULAS

Ripple, percent rms, sine-wave-shaped:

$$\text{ripple} = \frac{V_o \text{ (rms ripple)}}{V_O \text{ (dc FL)}} \times 100\%$$

$$= \frac{V_o \text{ (p-p ripple)}}{2.828 V_O \text{ (dc FL)}} \times 100\%$$

Ripple, percent rms, sawtooth-shaped:

$$\text{ripple} = \frac{V_o \text{ (rms ripple)}}{V_O \text{ (dc FL)}} \times 100\%$$

$$= \frac{V_o \text{ (p-p ripple)}}{3.47 V_O \text{ (dc FL)}} \times 100\%$$

Load regulation, percent, no-load (NL) to full-load (FL):

$$\text{load regulation} = \frac{V_O \text{ (FL)} - V_O \text{ (NL)}}{V_O \text{ (FL)}} \times 100\%$$

Line regulation, $V_{\text{line (max)}}$ to $V_{\text{line (min)}}$:

$$\text{line regulation} = \frac{\% \text{ load voltage change}}{\% \text{ line voltage change}} \times 100\%$$

$$= \frac{V_{\text{line(min)}}\left(V_{\text{load(max)}} - V_{\text{load(min)}}\right)}{V_{\text{load(min)}}\left(V_{\text{line(max)}} - V_{\text{line(min)}}\right)} \times 100\%$$

1.10 DECIBELS

Decibels express a power ratio, not an amount. They tell how many times more (positive db) or less (negative dB) but not how much. Decibels are logarithmic, not linear. 20 dB is not twice the power ratio of 10 dB.

$$\alpha_{dB} = 10 \log \frac{P_2}{P_1} \text{ dB}$$

Voltage is more easily measured than power, making it generally more convenient to use

$$\alpha_{dB} = 20 \log \frac{V_2}{V_1} \text{ dB}$$

The voltage-ratio formula is valid only if the two voltages appear across equal resistances. Where the resistances differ, calculate P_1 and P_2 and use the power-ratio formula.

Zero-dB standards:

Audio industry:
  0 dB = 1 mW in 600 Ω resistance
       = 0.7746 V rms across 600 Ω

Television industry:
  0 dB = 1mV rms across 75 Ω resistance
       = 1.333×10^{-8} W

Thus, a –52-dB microphone delivers a power P_2 of

$$P_2 = P_1 \times 10^{-52/10} = 1 \text{ mW} \times 6.3 \times 10^{-6}$$
$$= 6.3 \text{ nW}$$

Adding decibels is equivalent to multiplying gain factors.

Negative decibels represent loss factors (division):

  Voltage ratios: $2 \times 10 \times \frac{1}{4} = 5$
  Decibels: $6 + 20 - 12 = 14$

Decibel-to-ratio conversion (Table 1-2):

$$\frac{P_2}{P_1} = \log^{-1}\frac{\alpha_{dB}}{10} = 10^{(\alpha_{dB}/10)}$$

$$\frac{V_2}{V_1} = \log^{-1}\frac{\alpha_{dB}}{20} = 10^{(\alpha_{dB}/20)}$$

TABLE 1-2. Decibel to Voltage-Ratio to Power-Ratio Conversion

α_{dB}	$\frac{V_2}{V_1}$ $(R_1 = R_2)$	$\frac{P_2}{P_1}$	α_{dB}	$\frac{V_2}{V_1}$ $(R_1 = R_2)$	$\frac{P_2}{P_1}$	α_{db}	$\frac{V_2}{V_1}$ $(R_1 = R_2)$	$\frac{P_2}{P_1}$
0	1.000	1.000	16	6.310	39.81	46	199.5	39,810
0.5	1.059	1.122	17	7.079	50.12	48	251.2	63,100
1	1.122	1.259	18	7.943	63.10	50	316.2	1×10^5

2	1.259	19	1.585	8.913	52	79.43	398.1	1.58×10^5
3	1.413	20	1.995	10.00	54	100.0	501.2	2.51×10^5
4	1.585	22	2.512	12.59	56	158.5	631.0	3.98×10^5
5	1.778	24	3.162	15.85	58	251.2	794.3	6.31×10^5
6	1.995	26	3.981	19.95	60	398.1	1,000	1×10^6
7	2.239	28	5.012	25.12	62	631.0	1,259	1.58×10^6
8	2.512	30	6.310	31.62	64	1,000	1,585	2.51×10^6
9	2.818	32	7.943	39.81	66	1,585	1,995	3.98×10^6
10	3.162	34	10.00	50.12	68	2,512	2,512	6.31×10^6
11	3.548	36	12.589	63.10	70	3,981	3,162	1×10^7
12	3.981	38	15.85	79.43	75	6,310	5,623	3.16×10^7
13	4.467	40	19.95	100.0	80	10,000	10,000	1×10^8
14	5.012	42	25.12	125.9	85	15,850	17,780	3.16×10^8
15	5.623	44	31.62	158.5	90	25,120	31,620	1×10^9

1.11 INSTRUMENTATION DEFINITIONS

Precision. The degree to which a measurement is readable or is specified. May be indicated in units of measure (example: "to within ±10 mV") or in percent (example: "readable to within 0.5% of full scale").

Resolution. The smallest increment that will render one reading distinguishable from another.

Sensitivity. The ratio of output response to input stimulation. Often expressed as input required for full-scale (f.s.) output, or input required for minimum observable output.

Error. The difference between true and indicated values of the measured quantity. Often expressed as a percent of the true quantity or as a percent of full range of the instrument.

$$\epsilon = \frac{V_{indicated} - V_{true}}{V_{true}} \times 100\%$$

Accuracy. The degree to which the indicated value approaches the true value. Usually expressed by percent error (see above).

Linearity. The degree to which the graph of input stimulation vs. output response approaches a straight line. Expressed by percent *nonlinearity*, as illustrated in Fig. 1-14. *Normal linearity* has the end points of the ideal line and the actual curve coincident, as shown. *Zero-based linearity* has the zero points coincident while the slope of the ideal line is adjusted for lowest percent deviation from the actual curve. *Independent linearity* allows both end points to float as the ideal line is adjusted for lowest percent nonlinearity. The deviations are commonly expressed as a percent of full scale.

$$\text{Linearity} = \frac{\Delta V}{V_{max}} \times 100\%$$

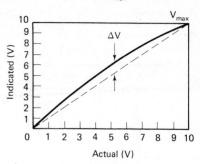

FIGURE 1-14

1.12 ATTENUATOR FORMULAS

Definitions:

$$\text{attenuation} = a = \frac{1}{A} = \frac{V_{in}}{V_{out}}$$

$Z_o = R_s = R_L$ (characteristic impedance, source resistance, and load resistance equal)

Pi-pad, O-pad [Fig. 1-15(a)]:

Design	*Analysis*

$$R_1 = Z_o \frac{a^2 - 1}{2a} \qquad Z_o = \frac{R_2}{\sqrt{\dfrac{R_1 + 2R_2}{R_1}}}$$

$$R_2 = Z_o \frac{a + 1}{a - 1} \qquad a = \frac{R_1 + R_2 \| R_1}{R_2 \| R_1}$$

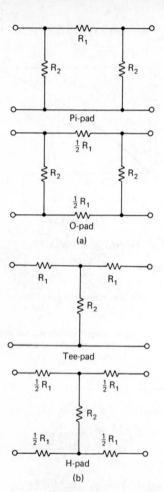

Pi-pad

O-pad

(a)

Tee-pad

H-pad

(b)

FIGURE 1-15

Tee-pad, H-pad [Fig. 1-15(b)]:

Design   *Analysis*

$$R_1 = Z_o \frac{a-1}{a+1} \qquad Z_o = R_1 \sqrt{\frac{R_1 + 2R_2}{R_1}}$$

$$R_2 = Z_o \frac{2a}{a^2 - 1} \qquad a = 1 + \frac{R_1 \sqrt{R_1{}^2 + 2R_1 R_2}}{R_2}$$

Padding a line reduces impedance mismatch when the line is open-circuited (Fig. 1-16). Z_o/R_{in} is similar for shorted line end.

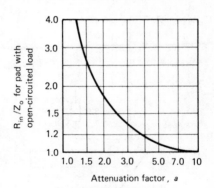

FIGURE 1-16

Frequency-compensated attenuator (Fig. 1-17):

$$a = \frac{V_{in}}{V_o} = \frac{R_1 + R_2}{R_1}$$

$$R_1 C_1 = R_2 C_2$$

Constant-resistance L-pad (Fig. 1-18): R_1 is linear, R_2 is logarithmic.

39

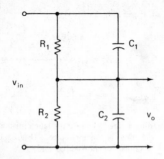

FIGURE 1-17 Compensated attenuator

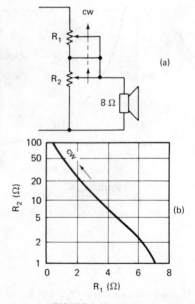

FIGURE 1-18 L-pad

1.13 BRIDGE CIRCUITS

Potentiometer bridge (Fig. 1-19):

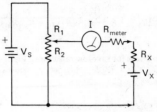

FIGURE 1-19

$$V_X = V_S \frac{R_2}{R_1 + R_2} \qquad I_m = \frac{\dfrac{V_S R_2}{R_1 + R_2} - V_X}{R_1 \| R_2 + R_m + R_x}$$

Wheatstone bridge (Fig. 1-20):

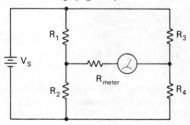

FIGURE 1-20

At Balance:

$$\frac{R_1}{R_2} = \frac{R_3}{R_4}$$

All resistors nominally equal R; one resistor changes by ΔR:

$$I_m = \frac{V_s \Delta R}{R(4R + R_m)}$$

Slightly unbalanced; resistance arms not equal; R_4 arm changes by ΔR:

$$I_m = \frac{V_S R_3 \Delta R}{(2R_3R_4 + R_3{}^2 + R_4{}^2)(R_1 \| R_2 + R_3 \| R_4 + R_m)}$$

Kelvin bridge for very low resistances (Fig. 1-21): Let $R_5/R_6 = R_1/R_2$; balance as Wheatstone bridge.

Series capacitance-comparison bridge for low D factors (Fig. 1-22):

Maxwell bridge for low-Q inductors (Fig. 1-23):

Hay bridge for high-Q inductors (Fig. 1-24):

Series-equivalent formulas for Hay bridge:

$$L_s = \frac{R_2 R_3 C_1}{1 + 2\pi f R_1{}^2 C_1{}^2}$$

$$R_s = \frac{2\pi f R_1 R_2 R_3 C_1}{1 + 2\pi f R_1{}^2 C_1{}^2}$$

Wien Bridge (Fig. 1-25): Nearly infinite attenuation at f_c. Six-dB attenuation at $2f_c$ and $1/2f_c$ requires very high R_L. Both sides isolated from ground. $V_o(\text{max}) = 1/3 V_s$.

Twin-tee notch filter (Fig. 1-26): Similar to Wein bridge, but grounded output. $V_o(\text{max}) = V_s$.

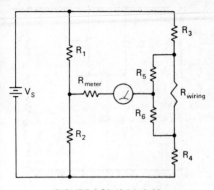

FIGURE 1-21 Kelvin bridge

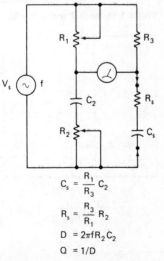

$$C_s = \frac{R_1}{R_3} C_2$$

$$R_s = \frac{R_3}{R_1} R_2$$

$$D = 2\pi f R_2 C_2$$

$$Q = 1/D$$

FIGURE 1-22 Capacitance comparison

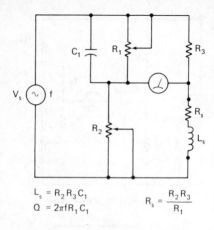

$$L_s = R_2 R_3 C_1$$
$$Q = 2\pi f R_1 C_1 \qquad R_s = \frac{R_2 R_3}{R_1}$$

FIGURE 1-23 Maxwell bridge (low Q)

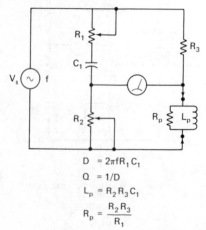

$$D = 2\pi f R_1 C_1$$
$$Q = 1/D$$
$$L_p = R_2 R_3 C_1$$
$$R_p = \frac{R_2 R_3}{R_1}$$

FIGURE 1-24 Hay bridge (high Q)

44

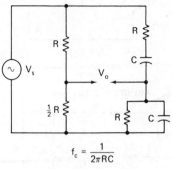

$$f_c = \frac{1}{2\pi RC}$$

FIGURE 1-25 Wien bridge

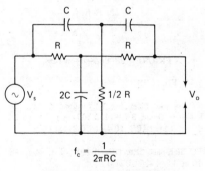

$$f_c = \frac{1}{2\pi RC}$$

FIGURE 1-26 Twin-tee notch filter

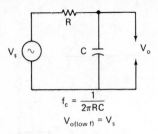

$$f_c = \frac{1}{2\pi RC}$$

$$V_{o(\text{low } f)} = V_s$$

FIGURE 1-27 Low-pass filter

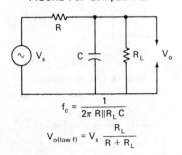

$$f_c = \frac{1}{2\pi \, R\|R_L \, C}$$

$$V_{o(\text{low } f)} = V_s \, \frac{R_L}{R + R_L}$$

FIGURE 1-28 Loaded low-pass

1.14 FILTERS

RC low-pass filter (Fig. 1-27 unloaded, Fig. 1-28 loaded): Down 3 dB ($\times$ 0.707) at f_c. Down 20 dB ($\times$ 0.1) at $10f_c$, 40 dB ($\times$ 0.01) at $100f_c$.

RC high-pass filter (Fig. 1-29): -3 dB at f_c, -6 dB at $0.5f_c$, -20 dB at $0.1f_c$.

Cascade *RC* high- or low-pass filters: Down 6 db at f_c for two-section if R_2 (second section) is $3R_1$ (first section). C_2 must then be $1/3C_1$. Down 10 dB at f_c for equal *R* and *C* values in both sections. Attenuation -12 dB at $2f_c$ (or

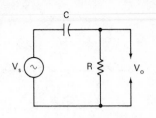

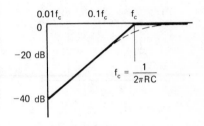

$$f_c = \frac{1}{2\pi RC}$$

FIGURE 1-29 High-pass filter

$1/2\ f_c$), –40 dB at $10\ f_c$ (or $1/10\ f_c$) for two sections.

Bandpass *RC* filter: Follow low-pass section with high-pass section. Keep f_{hi} at least $10f_{low}$. Let $R_1 = R_2$.

LC constant -*k* low- and high-pass filters (Fig. 1-30) have 3-dB attenuation at f_c and 18-dB attenuation for each factor of 2 away from f_c. Two, three, or four sections may be cascaded to achieve 36-dB, 54-dB, or 72-dB attenuation at a frequency factor of 2. Source and load resistances must equal Z_o. Impedance departs from Z_o in stopband and in passband near f_c, causing reflections and standing waves.

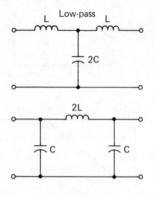

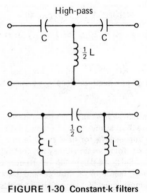

FIGURE 1-30 Constant-k filters

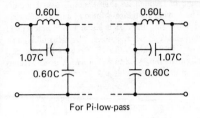

For Pi-low-pass

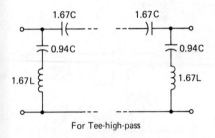

For Tee-high-pass

FIGURE 1-31 End sections for Fig. 1-30

$$
\begin{array}{cc}
Analysis & Design \\
Z_o = \sqrt{\dfrac{L}{C}} & L = \dfrac{Z_o}{2\pi f_c} \\[2ex]
f_c = \dfrac{1}{2\pi\sqrt{LC}} & C = \dfrac{1}{2\pi f_c Z_o}
\end{array}
$$

End sections (Fig. 1-31) keep Z constant in passband and place attenuation notch just outside passband.

Bandpass *LC* filters (Fig. 1-32) pass a much wider band of frequencies than a simple tuned circuit. The driving-source resistance r_g and the load resistance R_L must each equal the characteristic impedance Z_o of the filter. The ratio of high to low –3-dB frequencies (f_2/f_1) should generally be in the range 1.2 to 5.0 to avoid severe attenuation in the passband.

$$L_1 = \frac{Z_o}{\pi(f_2 - f_1)} \qquad C_1 = \frac{f_2 - f_1}{4\pi Z_o f_1 f_2}$$

$$L_2 = \frac{Z_o(f_2 - f_1)}{4\pi f_1 f_2} \qquad C_2 = \frac{1}{\pi Z_o (f_2 - f_1)}$$

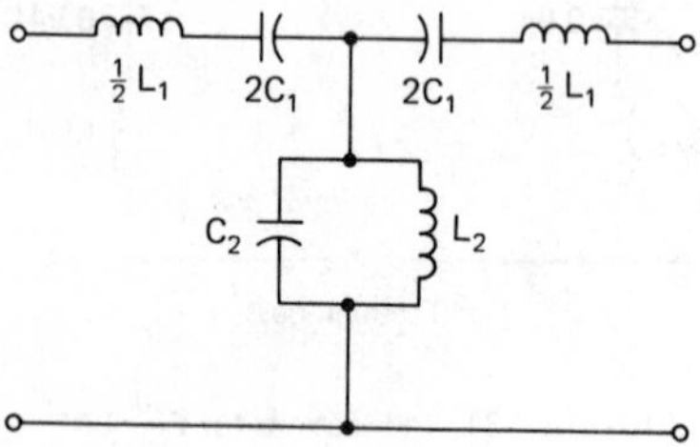

Tee bandpass

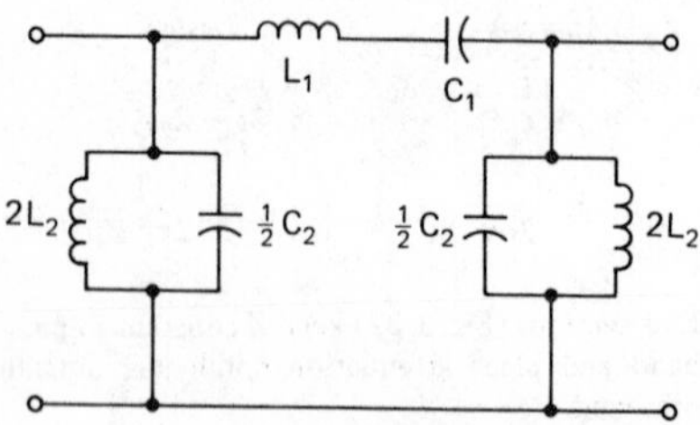

Pi bandpass

FIGURE 1-32

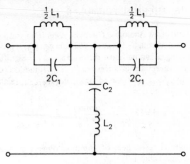

Tee bandstop

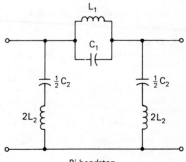

Pi bandstop

FIGURE 1-33

Bandstop *LC* filters (Fig. 1-33) eliminate a relatively broad range of frequencies while passing frequencies below f_1 or above f_2.

$$L_1 = \frac{Z_o(f_2 - f_1)}{\pi f_1 f_2} \qquad C_1 = \frac{1}{4\pi Z_o(f_2 - f_1)}$$

$$L_2 = \frac{Z_o}{4\pi(f_2 - f_1)} \qquad C_2 = \frac{f_2 - f_1}{\pi Z_o(f_1 f_2)}$$

1-15 THERMAL FORMULAS

Ohm's Law Analogy: Thermal power (P in watts) flows through thermal resistance (R_θ), producing a temperature differential (ΔT). Significant thermal resistances are $R_{\theta(J-C)}$ (junction to case), $R_{\theta(C-S)}$ (case to sink), and $R_{\theta(S-A)}$ (sink to ambient).

$$P = \frac{\Delta T}{R_\theta} = \frac{T_J(\text{max}) - T_A}{R_{\theta(J-C)} + R_{\theta(C-S)} + R_{\theta(S-A)}}$$

Thermal resistances of common objects in free air (see Table 1-3)

TABLE 1-3. Thermal Resistance of Common Objects

Transistor case	$R_{\theta(C-A)}$
TO-92 plastic	300
TO-18 mini TO-5	300
TO-5 standard	150
TO-60 stud	70
TO-66 mini TO-3	60
TO-220 tab	50
TO-3 std pwr	30
TO-36 1¼ in. round	25

Black metal plate, area one side (cm²)	$R_{\theta(S-A)}$
12	30
20	14
30	8
60	5
150	3

Temperature rise vs. resistance: The temperature rise of transformers, motors, or any device containing measurable metallic resistances (Tables 1-4 and 1-5) can be calculated by the resistance change from cold to hot:

$$\frac{R_H}{R_C} = K^{(T_H - T_C)} \qquad T_H - T_C = \frac{\log(R_H/R_C)}{\log K}$$

where R is resistance, T temperature in °C, K the temperature coefficient of resistance of the metal per °C, and H and C represent hot and cold, respectively.

TABLE 1-4. K (per °C) for Common Metals

Metal	$K/°C$
Aluminum	1.0040
Copper	1.0039
Gold	1.0034
Iron	1.0055
Nichrome	1.00017
Silver	1.0038
Tungsten	1.005

TABLE 1-5. Resistance vs. Temperature (°C) for Copper

R_H/R_C	$T_H - T_C$	R_H/R_C	$T_H - T_C$
1.04	10	1.37	80
1.08	20	1.48	100
1.12	30	1.60	120
1.17	40	1.72	140
1.21	50	1.86	160
1.26	60	2.02	180
1.31	70	2.18	200

1.16 Mathematical Formulas

Geometric Formulas

Definitions: A = area, a = altitude, b = base, c = hypotenuse, C = circumference, d = diameter, h = height, r = radius, s = side of a regular polygon, V = volume.

Circle: $C = 2\pi r = \pi d$; $\quad A = \pi r^2$

Sector (Fig. 1-34): $A = 1/2\, sr = \pi r^2\, \dfrac{\theta°}{360°}$

Segment: $A = \pi r^2\, \dfrac{\phi°}{360°} - 1/2\, b(r - h)$;

$$b = 2\sqrt{2hr - h^2}$$

Sphere: $A = 4\pi r^2 = \pi d^2$; $\quad V = 4/3\, \pi r^3$

Ellipse: $A = 1/4\, \pi d_1 d_2$, where d_1 and d_2 are the major and minor axes (diameters)

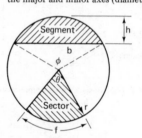

FIGURE 1-34

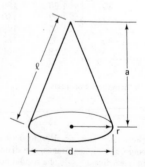

FIGURE 1-35

54

Cone (Fig. 1-35): $A = \pi r l = \pi r \sqrt{r^2 + a^2}$ (not including base disk);
$$V = 1/3\pi r^2 a$$

Torus [doughnut with circular cross section (Fig. 1-36)]: $A = \pi^2 dD = 4\pi^2 rR$;
$$V = 2.463 d^2 D = 2\pi^2 r^2 R$$

Regular pentagon: $A = 1.720 s^2$

Regular hexagon: $A = 2.598 s^2$

Regular octagon: $A = 4.828 s^2$

Regular polygon of n sides: $A = \dfrac{s^2}{4 \tan\dfrac{180°}{n}}$

Triangle, right (Fig. 1-37): $A = 1/2\, ba$;
$$c^2 = a^2 + b^2$$

$$\sin A = \frac{a}{c} \qquad \csc A = \frac{c}{a}$$

$$\cos A = \frac{b}{c} \qquad \sec A = \frac{c}{b}$$

$$\tan A = \frac{a}{b} \qquad \cot A = \frac{b}{a}$$

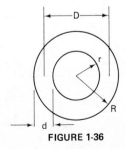

FIGURE 1-36

55

Triangle, oblique (Fig. 1-38):

$$A = 1/2\,bh$$

$$A = \sqrt{m(m-a)(m-b)(m-c)},$$

where $m = \dfrac{a+b+c}{2}$

$$\dfrac{a}{\sin A} = \dfrac{b}{\sin B} = \dfrac{c}{\sin C} \qquad \text{(sine law)}$$

$$a^2 = b^2 + c^2 - 2bc \cos A \qquad \text{(cosine law)}$$

FIGURE 1-37

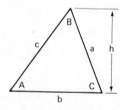

FIGURE 1-38

Determinants for the solution of n linear equations in n unknowns:

Standard form of the equations (in two unknowns):

$a_1 x + b_1 y = k_1$

$a_2 x + b_2 y = k_2$

$$x = \frac{\begin{vmatrix} k_1 & b_1 \\ k_2 & b_2 \end{vmatrix}}{\begin{vmatrix} a_1 & b_1 \\ a_2 & b_2 \end{vmatrix}} = \frac{k_1 b_2 - b_1 k_2}{a_1 b_2 - b_1 a_2}$$

$$y = \frac{\begin{vmatrix} a_1 & k_1 \\ a_2 & k_2 \end{vmatrix}}{\begin{vmatrix} a_1 & b_1 \\ a_2 & b_2 \end{vmatrix}} = \frac{a_1 k_2 - k_1 a_2}{a_1 b_2 - b_1 a_2}$$

Third-order determinant (technique not valid for higher orders):

$$\begin{vmatrix} a_1 & b_1 & c_1 \\ a_2 & b_2 & c_2 \\ a_3 & b_3 & c_3 \end{vmatrix} \begin{matrix} a_1 & b_1 \\ a_2 & b_2 \\ a_3 & b_3 \end{matrix} = \begin{matrix} a_1 b_2 c_3 + b_1 c_2 a_3 + c_1 a_2 b_3 \\ -c_1 b_2 a_3 - a_1 c_2 b_3 - b_1 a_2 c_3 \end{matrix}$$

Fourth-order determinant:

$$D = a_1(b_2 c_3 d_4 + c_2 d_3 b_4 + d_2 b_3 c_4$$
$$- d_2 c_3 b_4 - b_2 d_3 c_4 - c_2 b_3 d_4)$$

$$- a_2(b_1 c_3 d_4 + c_1 d_3 b_4 + d_1 b_3 c_4$$
$$- d_1 c_3 b_4 - b_1 d_3 c_4 - c_1 b_3 d_4)$$

$$+ a_3(b_1 c_2 d_4 + c_1 d_2 b_4 + d_1 b_2 c_4$$
$$- d_1 c_2 b_4 - b_1 d_2 c_4 - c_1 b_2 d_4)$$

$$- a_4(b_1 c_2 d_3 + c_1 d_2 b_3 + d_1 b_2 c_3$$
$$- d_1 c_2 b_3 - b_1 d_2 c_3 - c_1 b_2 d_3)$$

Determinant simplification (example in third order):

Multiply column 2 by 3 and subtract from column 1:

$$\begin{vmatrix} 8 & 2 & 5 \\ 9 & 7 & 4 \\ 3 & 1 & 6 \end{vmatrix} = \begin{vmatrix} 8-6 & 2 & 5 \\ 9-21 & 7 & 4 \\ 3-3 & 1 & 6 \end{vmatrix} = \begin{vmatrix} 2 & 2 & 5 \\ -12 & 7 & 4 \\ 0 & 1 & 6 \end{vmatrix}$$

Multiply row 1 by 6 and add to row 2:

$$
\begin{vmatrix} 2 & 2 & 5 \\ -12 & 7 & 4 \\ 0 & 1 & 6 \end{vmatrix} = \begin{vmatrix} 2 & 2 & 5 \\ -12+12 & 7+12 & 4+30 \\ 0 & 1 & 6 \end{vmatrix} = \begin{vmatrix} 2 & 2 & 5 \\ 0 & 19 & 34 \\ 0 & 1 & 6 \end{vmatrix}
$$

Quadratic Formula

Standard form: $ax^2 + bx + c = 0$

Solution: $x = \dfrac{-b \pm \sqrt{b^2 - 4ac}}{2a}$

Boolean Algebra Postulates and Theorems

1. $0 \cdot 0 = 0$ $0 + 0 = 0$
2. $0 \cdot 1 = 0$ $0 + 1 = 1$
3. $1 \cdot 1 = 1$ $1 + 1 = 1$
4. $\bar{1} = 0$ $\bar{0} = 1$
5. if $A = 0, \bar{A} = 1$ if $A = 1, \bar{A} = 0$
6. $A \cdot 0 = 0$ $A + 0 = A$
7. $A \cdot 1 = A$ $A + 1 = 1$
8. $A \cdot A = A$ $A + A = A$
9. $A \cdot \bar{A} = 0$ $A + \bar{A} = 1$
10. $\bar{\bar{A}} = A$ $A = \bar{\bar{A}}$
11. $A \cdot B = B \cdot A$ $A + B = B + A$
    (commutative)
12. $A \cdot (B \cdot C) = A \cdot B \cdot (C)$ $A + (B+C) = (A+B) + C$
    (associative)
13. $A \cdot (B+C) = A \cdot B + A \cdot C$ $A + B \cdot C = (A+B) \cdot (A+C)$
    (distributive)
14. $A \cdot (A + B) = A$ $A + \bar{A} \cdot B = A$
15. $A \cdot (A + B) = A \cdot B$ $A + \bar{A} \cdot B = A + B$
    (absorption)
16. $\overline{A \cdot B \cdot C} = \bar{A} + \bar{B} + \bar{C}$ $\overline{A + B + C} = \bar{A} \cdot \bar{B} \cdot \bar{C}$
    (de Morgan)

2

Component Data
and Characteristics

2.1 WIRE AND CABLE

Copper wire: allowable currents and characteristics (See Tables 2-1 and 2-2)

Insulation types:

Rubber has maximum flexibility but is susceptible to abrasion and becomes brittle with age and exposure to sunlight. Silicon conditioning compounds help prolong life.

Polyvinyl Chloride (PVC) is popular for hookup wire and other indoor applications where temperatures do not exceed 70°C. It has good resistance to water, oil, and most chemical contaminants, and remains quite stable with age. It should not be used for capacitor dielectric because of energy absorption.

Teflon is less flexible than PVC but is also resistant to sunlight, moisture, and most chemicals. It may be used at temperatures up to 260°C. Fumes from burned Teflon may cause symptoms of influenza. Particular care should be taken to avoid contamination of smoking tobacco with Teflon particles.

Polyethylene has good resistance to environmental conditions and retains its flexibility at low temperatures. It has excellent dielectric properties and is widely used in coaxial cables.

Enamel and such trade-name improvements upon it as *Formvar* and *Nyclad* are used where wire thickness must be kept to a minimum and exposure to flexing and abrasion will not be a problem. The primary application is in winding coils, from dc to VHF, Temperatures to 105°C can be tolerated.

TABLE 2-1. Allowable currents (amperes) for Copper Wire, Based on 30°C Ambient, 100°C Final Temperature

AWG Size	Single Wire in Free Air	Bundled Wires, Confined
6	95	55
8	62	39
10	50	31
12	40	23
14	32	17
16	22	13
18	16	10
20	11	7.5
22	7	5
24	3.5	2.1
26	2.2	1.5
28	1.4	0.8
30	0.8	0.5
32	0.5	0.3

Note: Values are given only as a guide. Degree of air movement and confinement will greatly affect results.

TABLE 2-2. Characteristics of Copper Wire at 20°C

Notes:

[1] Each wire size number increase represents a factor-of-1.26 resistance increase over the previous size. Moving three gage sizes higher doubles the resistance.

[2] To convert table values to diameter in mm, multiply by 0.0254.

[3] To convert table values to area in mm$^2 \times 10^{-6}$, multiply by 645.

[4] To convert table values to Ω/m, multiply by 0.0394.

[5] To convert table values to m/kg, multiply by 0.672.

[6] Turns per in.^{2} cross-sectional area based on machine winding in a 1 in. $\times$ 1 in. channel. To convert to turns per cm^2, multiply by 0.155.

[7] Other strandings often available. Note that total cross-sectional area of stranded wire may vary as much as +20, −6% from solid-wire area.

Size AWG[1]	Diameter[2] (mils)	Area[3] cmil	Area[3] in.$^2 \times 10^{-6}$	Ohms per 1000 ft[4]	Feet per Pound[5]	Turns per sq. in.[6]	Typical Strandings[7]
0000	460.0	211 600	166 200	0.04901	1.5	—	2104/30
000	409.6	167 800	131 800	1.06182	1.9	—	1661/30

61

Size AWG [1]	Diameter [2] (mils)	Area [3] cmil	Area [3] in.² × 10⁻⁶	Ohms per 1000 ft [4]	Feet per Pound [5]	Turns per sq. in. [6]	Typical Strandings [7]
00	364.8	133 100	104 500	0.07793	2.4	—	1330/30
0	324.9	105 600	82 910	0.09825	3.1	—	1045/30
1	289.3	83 690	65 730	0.1239	3.9	—	817/30
2	257.6	66 360	52 120	0.1563	4.9	—	665/30
3	229.4	52 620	41 330	0.1971	6.2	—	–
4	204.3	41 740	32 780	0.2485	7.8	—	133/25
5	181.9	33 090	25 990	0.3134	10	—	–
6	162.0	26 240	20 610	0.3952	12	—	133/27
7	144.3	20 820	16 350	0.4981	16	—	–
8	128.5	16 510	12 970	0.6281	20	—	133/29, 168/30
9	114.4	13 090	10 280	0.7925	25	—	–
10	101.9	10 380	8 155	0.9988	32	88	19/22, 105/30
11	90.7	8 226	6 461	1.260	40	112	–
12	80.8	6 529	5 128	1.590	50	140	19/25, 65/30
13	72.0	5 184	4 072	2.00	63	176	–

14	64.1	4 109	3 227	2.52	80	220	19/27, 41/30
15	57.1	3 260	2 561	3.18	101	260	—
16	50.8	2 581	2 027	4.02	128	320	19/29, 26/30
17	45.3	2 052	1 612	5.05	160	400	—
18	40.3	1 624	1 276	6.39	203	500	16/30, 65/36
19	35.9	1 289	1 012	8.05	256	630	—
20	32.0	1 024	804	10.1	323	790	41/36, 10/30
21	28.5	812	638	12.8	401	990	19/33
22	25.3	640	503	16.2	514	1 240	26/36, 7/30
23	22.6	511	401	20.3	644	1 530	10/32, 16/34
24	20.1	404	317	25.7	818	1 890	16/36, 7/32
25	17.9	320	252	32.4	1 031	2 300	8/34, 12/36
26	15.9	253	199	41.0	1 330	2 900	10/36, 26/40
27	14.2	202	158	51.4	1 639	3 700	8/36, 7/35
28	12.6	159	125	65.3	2 067	4 581	6/36, 7/36
29	11.3	128	100	81.2	2 607	5 600	5/36
30	10.0	100	78.5	104	3 287	7 000	7/38, 4/36
31	8.9	79.2	62.2	131	4 145	8 400	3/35, 6/38

Size AWG[1]	Diameter[2] (mils)	Area[3] cmil	Area[3] in.² × 10⁻⁶	Ohms per 1000 ft[4]	Feet per Pound[5]	Turns per sq. in.[6]	Typical Strandings[7]
32	8.0	64.0	50.3	162	5 230	10 500	4/38, 7/40
33	7.1	50.4	39.6	206	6 591	13 100	—
34	6.3	39.7	31.2	261	8 311	16 800	—
35	5.6	31.4	24.6	331	10 480	21 000	—
36	5.0	25.0	19.6	415	13 210	26 000	—
37	4.5	20.2	15.9	512	16 660	31 000	—
38	4.0	16.0	12.6	648	21 010	39 000	—
39	3.5	12.2	9.62	847	26 500	53 000	—
40	3.1	9.61	7.55	1 080	34 400	65 000	—
41	2.8	7.84	6.16	1 320	42 000	—	—
42	2.5	6.25	4.91	1 660	53 000	—	—
43	2.2	4.84	3.80	2 140	68 000	—	—
44	2.0	4.00	3.14	2 590	82 000	—	—

Conductivity and temperature coefficient of various metals (see Table 2-3)

Note: To determine the resistivity of conductors that are not wire-shaped, use

$$R = \frac{\rho\, l}{A}$$

where R is in Ω, ρ is in $\Omega \cdot cm \times 10^{-6}$ from the table, l is length in cm, and A is cross-sectional area in cm^2.

Popular Coaxial Connectors

Plug (male): center-conductor pin; outer conductor slips over receptacle.

Jack or receptacle (female): center conductor receives pin.

	BNC (UG–XXX)	UHF
Line plug	88*, 260	PL-259
Line jack	89*, 261	
Chassis jack (4-screw mount)	290, 447	SO-239
Chassis jack (1 hole & nut)	625, 912, 1094	
Double jack, line	914	PL-258
Double jack, chassis	414	UG-363
Double plug, line	491	
Tee; 2 jack, 1 plug	274	M-358

* For RG-58 size cable or smaller only.

Characteristics of coaxial cables (see Table 2-4)

Note: Attenuation data are for new cable at 20° C. Moisture intrusion, heat, and age can greatly increase attenuation. Velocity factor for all cables in table is approximately $0.66c$ or 200×10^6 m/s. Velocity factor for 300-Ω twin lead is about $0.8c$, and for open-wire line, about 0.97.

TABLE 2-3. Conductivity and Temperature Coefficient of Various Conductors at 20°C

Material	Resistivity Relative to Copper	Resistivity (Ω·cmil/ft)	Resistivity, ρ (μΩ·cm)	Temperature Coefficient per °C at 20°C
Aluminum	1.64	17	2.83	+0.0040
Brass	3.58	37	7.0	+0.0015
Copper	1.00	10.37	1.724	+0.0039
Gold	1.42	14.7	2.44	+0.0034
Iron	5.59	58	9.64	+0.0055
Lead	11.86	123	20.4	+0.0039
Mercury	100	—	172	—
Nichrome	65.09	675	112	+0.0017
Silver	0.945	9.80	1.63	+0.0038
Steel (soft)	9.24	95.8	15.9	+0.0016
Tin	6.68	69.3	11.5	+0.0042
Tungsten	3.20	33.2	5.52	+0.005

TABLE 2-4. Typical Characteristics of Popular Coaxial Cables

Type RGXX/U	Z_o (Ω)	O.D. (in.)	Weight (lb/100 ft)	V_{max} (V_{rms})	Capacitance (pF/ft)	Attenuation (dB/100 ft) at f (MHz)						
						1	10	50	100	200	400	1000
8	52	0.40	11	5000	29.5	0.16	0.55	1.3	2.0	3.5	4.5	8.5
9	51	0.42	16	5000	30	0.12	0.47	1.4	1.9	2.9	4.4	8.0
11	75	0.41	10	5000	20.5	0.18	0.62	1.6	2.2	3.3	4.7	8.0
58	52	0.19	2.5	1900	30	0.40	1.3	3.2	5.0	8.0	12	22
59	73	0.24	3.2	2300	21	0.30	1.1	2.4	3.8	4.9	8.5	14
62	93	0.24	3.9	750	13.5	0.25	0.83	1.8	2.7	4.0	5.6	9.0
174	50	0.10	—	—	30	2.3	3.9	6.6	8.9	12	18	30
178	50	0.08	—	1000	29	2.6	5.6	10	14	20	28	46
179	75	0.11	1	1200	20	3.0	5.3	8.1	10	13	16	24
180	95	0.15	1.5	1500	15.5	2.4	3.3	4.6	5.7	7.6	11	17
213	50	0.41	12	5000	29.5	0.16	0.55	1.3	2.0	3.5	4.5	8.5
214	50	0.43	16	5000	30	0.12	0.47	1.4	1.9	2.9	4.4	8.0
223	50	0.22	3.6	1900	28.5	0.36	1.2	3.2	4.8	7.0	10	17

Color codes (see Table 2-5)

TABLE 2-5. Color Code for Chassis Wiring

Color	Use
Black	Grounds
Brown	Heaters or filaments
Red	Power supply, main positive
Orange	Secondary positive supplies; screen grids
Yellow	Emitters; cathodes
Green	Transistor bases; control grids
Blue	Collectors; plates (anodes)
Violet	Power supply, minus
Gray	Ac power lines
White	Miscellaneous

2.2 RESISTORS AND CAPACITORS

Standard values (see Tables 2-6 and 2-7)

TABLE 2-6. Standard Values in 20%, 10%, and 5% Components

10% Values (20% Values in Italic)	5% Values (in addition to 10% Values)
1.0	1.1
1.2	1.3
1.5	1.6
1.8	2.0
2.2	2.4
2.7	3.0
3.3	3.6
3.9	4.3
4.7	5.1
5.6	6.2
6.8	7.5
8.2	9.1
10	11

TABLE 2-7. Standard Decade of Values for 1% Components

100	147	215	316	464	681
102	150	221	324	475	698
105	154	226	332	487	715
107	158	232	340	499	732
110	162	237	348	511	750
113	165	243	357	523	768
115	169	249	365	536	787
118	174	255	374	549	806
121	178	261	383	562	825
124	182	267	392	576	845
127	187	274	402	590	866
130	191	280	412	604	887
133	196	287	422	619	909
137	200	294	432	634	931
140	205	301	442	649	953
143	210	309	453	665	976

Color codes for resistors and capacitors (see Table 2-8)

Notes to table (following page)

*Guaranteed minimum value, -0 + 100%. (4th column of table).
†White fifth band on composition resistors indicates solderable terminal. (6th column of table).

TABLE 2-8. Color Codes for Resistors and Capacitors

| Color | Significant Figures | Multiplier | Tolerance | | | Failure Rate Per 1000 h (%) | Temperature Coefficient (ppm/°C) | Dc Working Voltage |
			R (%)	C, 10 pF or Less (pF)	C, Over 10 pF (%)			
Black	0	1	±20	±2	±20	—	0	—
Brown	1	10	±1	±0.1	±1	1.0	−33	100
Red	2	100	±2	—	±2	0.1	−75	—
Orange	3	1 000	±3	±0.25	±2.5	0.01	−150	300
Yellow	4	10 000	GMV*	—	—	0.001	−220	—
Green	5	100 000	±5	±0.5	±5	—	−330	500
Blue	6	1 000 000	—	—	—	—	−470	—
Violet	7	10 000 000	—	—	—	—	−750	—
Gray	8	0.01	—	±0.25	—	—	+30	—
White	9	0.1	—	±1	±10	†	+500	—
Gold	—	0.1	±5	—	±5	—	+100	1000
Silver	—	0.01	±10	—	±10	—	Bypass	—
No color	—	—	±20	—	±20	—	—	—

70

Resistor and capacitor marking systems (see Fig. 2-1)

COMPOSITION, FILM AND WIREWOUND RESISTORS

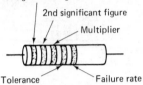

1st significant figure*

2nd significant figure

Multiplier

Tolerance   Failure rate

*Double-width band
indicates wirewound type.

PRECISION FILM RESISTOR

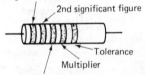

1st significant figure

2nd significant figure

Tolerance

Multiplier

3rd significant figure

BODY-END-DOT

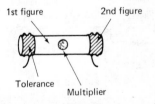

1st figure   2nd figure

Tolerance   Multiplier

FIGURE 2-1

MOLDED CERAMIC CAPACITOR

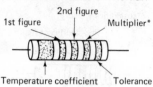

1st figure
2nd figure
Multiplier*
Temperature coefficient
Tolerance

*Values given in picofarads

MOLDED TUBULAR CAPACITOR

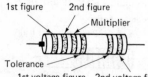

1st figure
2nd figure
Multiplier
Tolerance
1st voltage figure
2nd voltage figure

Add two zeros to voltage (Blk Red = 200 V)

MINIATURE
RESISTOR

1st figure
2nd figure
Multiplier

3-DOT DISC CAPACITOR

2nd figure
1st figure
Multiplier

FIGURE 2-1 (Cont.)

72

CURRENT MICA CAPACITOR CODE

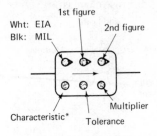

Wht: EIA
Blk: MIL

1st figure
2nd figure

Characteristic*
Tolerance
Multiplier

FRONT

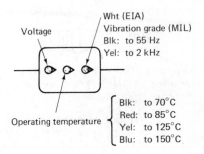

Voltage

Wht (EIA)
Vibration grade (MIL)
Blk: to 55 Hz
Yel: to 2 kHz

Operating temperature

Blk: to 70°C
Red: to 85°C
Yel: to 125°C
Blu: to 150°C

BACK (optional)

*Characteristic involves Q, temperature
coefficient and reliability specifications,
and ranges from Blk (least stringent) to
Grn (most stringent)

FIGURE 2-1 (Cont.)

5-DOT DISC CAPACITOR

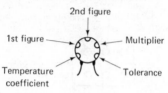

TUBULAR CERAMIC CAPACITOR

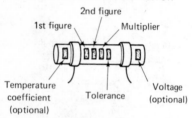

FIGURE 2-1 (Cont.)

EIA code for ceramic capacitors: In the EIA code for ceramic capacitors (Table 2-9), values are specified in pF by whole numbers or in μF by decimal numbers, followed by a letter indicating tolerance. Low-temperature limit, high-temperature limit, and maximum capacitance change over this temperature range are given, respectively, by a three-character code consisting of a letter, a number, and a letter. For temperature-compensating capacitors this three-character code is replaced by the letter N (negative) or P (positive) and a number representing capacitance change in parts per million per degree Celsius. The letters NPO indicate zero temperature coefficient. The dc operating voltage limit may also be given.

0.001 K Y5D 200V
value tol. temp. voltage

68 J N750
value tol. temp.

0.001μF, $\pm$ 10%; $\pm$3.3%
change from $-30°$C
to $+85°$C

68 pF, $\pm$5%;
temperature
coefficient,
-750 ppm/$°$C

TABLE 2-9. EIA Code for Ceramic Capacitors

Tolerance (%)		Temperature Characteristics		$\triangle C_{min}$ to C_{max} (%)	
		T_{min} (°C)			
		X	-55	A	± 1
F	± 1	Y	-30	B	± 1.5
G	± 2	Z	$+10$	C	± 2.2
H	± 3			D	± 3.3
J	± 5	T_{max} (°C)		E	± 4.7
K	± 10	5	$+85$	F	± 7.5
M	± 20	7	$+125$	P	± 10
Z	$+80, -20$			R	± 15
P	$+100, -0$ (also called GMV:			S	± 22
	GMV: guaranteed			T	$+22, -33$
	minimum value)			U	$\pm 22, -56$
				V	$+22, -82$

MIL-STD resistor designations

TABLE 2-10. Example of MIL-STD Resistor Designations

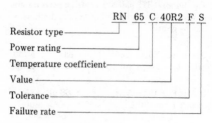

```
                              RN  65  C  40R2  F  S
Resistor type ──────────┘     │   │   │    │   │  │
Power rating ─────────────────┘   │   │    │   │  │
Temperature coefficient ──────────┘   │    │   │  │
Value ────────────────────────────────┘    │   │  │
Tolerance ─────────────────────────────────────┘  │
Failure rate ─────────────────────────────────────┘
```

The resistor described in Table 2-10 is a metal film, ¼-W type, with a temperature coefficient of ± 50 ppm/°C, a value of 40.2Ω, $\pm 1\%$, with a failure certified to be not greater than 0.001%/ 1000 h.

- *Type and power rating:*

- RA: variable, wirewound, precision

  code: 20 25 30
  power (W): 2   3   4

- RB: fixed, wirewound precision

code:	08	16	17	18	19	52	53
power (W):	$\frac{1}{2}$	$\frac{2}{3}$	1	$1\frac{1}{2}$	2	1	$\frac{1}{2}$

code:	55	56	57	58	70	71
power (W):	$\frac{1}{4}$	$\frac{1}{8}$	1	2	$\frac{1}{2}$	$\frac{1}{4}$

- RC: fixed, composition

code:	05	07	09	20	30	32	41	42
power (W):	$\frac{1}{8}$	$\frac{1}{4}$	$\frac{1}{2}$	$\frac{1}{2}$	1	1	2	2

- RD: power, film, noninductive

code:	31	33	35	37	39	60	65	70
power (W):	7	13	25	55	115	1	2	4

- RE: power, wirewound, with heat sink
  code: 60   65   70   75   77   80
  power (W): $7\frac{1}{2}$ 20   25   50 100 250

- RL: fixed, film
  code: 07   20   32   42
  power (W): $\frac{1}{4}$ $\frac{1}{2}$ 1   2

- RN: fixed, film, high stability
  code: 05   50   55   60   65   70   75
  power (W): $\frac{1}{8}$ $\frac{1}{20}$ $\frac{1}{10}$ $\frac{1}{8}$ $\frac{1}{4}$ $\frac{1}{2}$ 1

- RP: variable, power, wirewound
  code: 10   11   15   16   20   25   30
  power (W): 25   12   50   25   75 100 150

  code: 35   40   45   50   55
  power (W): 225 300 500 750 1000

- RV: variable, composition
  code: 01   04   05   06
  power (W): $\frac{1}{4}$ 2 $\frac{1}{2}$ $\frac{1}{3}$

- RW: fixed, power, wirewound
  code: 55   56   67   68   69   70   74
  power (W): 5   10   5   10 $2\frac{1}{2}$ 1   5

  code: 78   79   80   81
  power (W): 10   3 $2\frac{1}{4}$ 1

Note: The letter R following any of the two-letter codes shown above indicates that the resistor has a reliability level established under appropriate military specification.

- *Temperature coefficient* (this position used alternately to specify temperature limit or lead sturcture):

  J, E = ±25; H, C = ±50; K, O = ±100,

all in parts per million per degree Celsius (ppm/°C).

- *Value:* three (or four) digits; the first two (or three) are significant; the last one gives the number of zeros to be added. The letter R may be substituted for one of the numbers to indicate the position of a decimal point, in which case the numbers following are significant figures rather than multipliers.

- *Tolerance:* F = ±1%, G = ±2%, J = ±5%, K = ±10%

- *Failure rate:* M = 1%, P = 0.1%, R = 0.01%, S = 0.001%, all per 1000 h of operation under specified conditions.

TABLE 2-11. Average Dielectric Constants

Material	K	Material	K
Vacuum	1.0000	Paper (paraffin-	
Air	1.0006	impregnated)	3–4
Wood (dry)	1.5–5	Rubber	3–5
Paper	2–3	Celluloid	4
Oil	2–4	Quartz	4–5
Teflon	2.1	Formica	4.7
Vaseline	2.16	Mica	4.5–8
Polyethylene	2.3	Glass (Pyrex)	4.8
Lucite	2.5	Steatite (ceramic)	5–6
Paraffin	2.5	Porcelain	5–6
Wax	2.6	Glass (window)	7–8
Polystyrene	2.6	Water (distilled)	80
Plexiglass	2.8	Ceramic (high-K)	to 7000

Capacitance calculation, neglecting end effects: Actual capacitance will be slightly less, the amount of error becoming significant to the extent that d is an appreciable fraction of the perimeter of A.

$$C = 8.85 \times 10^{-10} \frac{KA}{d} \qquad \text{(metric)}$$

where C is in farads, A the area between plates in cm^2, d the plate spacing in cm, and K the dielectric constant (Table 2-11).

2.3 INDUCTORS AND TRANSFORMERS

Inductors

Solenoid-wound coils (cylindrical shape):

$$L_{\mu H} = \frac{d^2 N^2}{46d + 101l} \qquad \text{(metric; } d \text{ and } l \text{ in cm)}$$

To determine number of turns required on a close-wound single-layer coil, given the wire thickness t (cm), coil diameter d (cm), and required inductance L (μH):

$$N = \frac{\sqrt{(50.5Lt)^2 + 46d^3\,L} + 50.5Lt}{d^2}$$

Magnetic-core coils—toroid, pot core, C-I core, or E-I core (Table 2-12):

$$L_{\mu H} = 0.012N^2\frac{\mu A}{l_c} \qquad \text{(metric, no air gap)}$$

where N is the number of turns, A the effective cross-sectional core area in cm^2, l_c the magnetic path length in cm, and μ the magnetic permeability of the core.

$$L_{\mu H} = 0.012N^2\,\frac{A}{l_g + l_c/\mu} \qquad \text{(metric, air gap length } l_g \text{)}$$

Because μ varies considerably with magnetizing force for most core materials, it is common practice to leave an air gap $l_g > l_c/\mu$ to linearize the inductor. For example, 200 turns on a 10-cm silicon iron core with a 1-cm cross section has an inductance of 19.2 mH at a low signal level and 1.92 H at a high signal level. If an air gap of 0.1 cm is introduced in the core, the inductor responds in a more nearly linear manner to different signal levels: 3.84 mH at low levels and 4.79 mH at high levels.

Maximum ac voltage across a magnetic core coil:

$$V_{rms(max)} = 4.4 \times 10^{-4} \, fNAB_{sat}$$

where f is in Hz, N is number of turns, A is magnetic-core cross-sectional area in cm, and B_{sat} is in teslas. Where inductance is known but N is not, this reduces to

$$V_{rms(max)} = 4.0 B_{sat} f\sqrt{Ll_e A}$$

where L is in henries and l_e is the effective magnetic path length in cm:

$$l_e = l_g + \frac{l_c}{\mu}$$

Maximum dc current in a magnetic-core coil:

$$I_{sat} = \frac{B_{sat} l_e}{1.26 \times 10^{-4} N}$$

where I is in amperes and the other terms are as defined above. Where L is known but N is not,

$$I_{sat} = 0.87 B_{sat} \sqrt{\frac{Al_c}{L}}$$

TABLE 2.12. Permeability and Saturation of Magnetic Core Materials

Material	Relative Permeability (Air = 1.00)		B_{sat} (teslas)
	μ_i (Low Level)	μ_m (High Level)	
Silicon iron	400	40 000	1.5
Iron/nickel alloy	3 000	20 000	1.0
Powdered iron (typical)	125	127	0.3
Ferrites 3B7, 3C8, TI, W-03	2 300	1 900	0.4
Ferrite 3E2A	5 000	1 800	0.3
Ferrite 3E3	12 000	1 900	0.4
Ferrite 3D3	750	1 500	0.3
Ferrite 4C4	125	600	0.2
Ferrite 1Z2	15	—	—

Pulse Transformers: When a dc voltage is applied across an inductor, the current rises to a final value V_p/R, as shown in Fig. 2-2. However, the magnetic core may saturate before this current is reached, resulting in a sudden loss of inductor and transformer action. The product of source voltage and time to saturation is constant for any transformer and is called the volt-microsecond limit of a pulse transformer.

$$V_p t = I_{sat} L$$

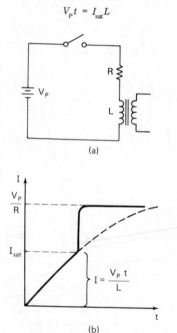

(a)

(b)

FIGURE 2-2 Pulse-transformer saturation

Sine-wave ac across an inductor may also cause I_{sat} to be reached if the voltage is high enough and the frequency is low enough to permit this peak current. The ratio of rms voltage to frequency is constant for any transformer and is called the volt-per-hertz limit.

$$V_{rms} = 4.4 I_{sat} L$$

Sometimes it is easier to test one of these limits than the other, making conversion to the other limit desirable:

$$V_{rms} = 4.4 V_p t$$

Transformer color codes (see Table 2-13)

TABLE 2-13. Transformer Color Codes

Power transformers

Black	Primary start
Black/yellow	Primary tap
Black/red	Primary finish
Red	High/voltage
Red/yellow	High-voltage tap
Green	Low-voltage No. 1
Green/yellow	Low-voltage No. 1 tap
Brown	Low-voltage No. 2
Gray	Low-voltage No. 3
Yellow	Rectifier filament

Audio and IF transformers

Blue	Primary signal source (plate or collector); finish
Red	Primary power-supply feed; center tap
Brown	Primary signal source; start
Green	Secondary signal out; finish
Black	Secondary ground return; center tap
Yellow	Secondary signal out; start

Stereo phono leads

Red	Right channel high
Green	Right channel low
White	Left channel high
Blue	Left channel low
Black	Ground

Stereo headphone plugs

Tip	Right channel
Ring	Left channel
Barrel	Common

2.4 SEMICONDUCTOR DEVICES

Diode markings (see Fig. 2-3)

Standard types of semiconductors. Table 2-14 covers the majority of discrete semiconductor applications with 22 inexpensive and widely available types.

Transistor case dimensions (see Fig. 2-4)

Pin connections for most-popular integrated circuits (see Fig. 2-5)

Notes:

1. The suffix N is used to identify plastic dual-in-line (DIP) packages having 8 to 28 pins. All diagrams are for N packages. Suffix J identifies ceramic DIP. Suffixes T and W identify flat packs.

2. Pinout for the 74L00 series often differs from the pinouts given.

3. TTL logic-level limits for all types; 54/74/L/ and LS series:
High input: 2.0 V and above.
Low input: 0.7 V* and below.
High output: 2.4 V or above.
Low output: 0.5V and below.
   * Some Schmitt triggers require 0.5 V

TABLE 2-14. Recommended Standard Semiconductor Types

Bipolar transistors	*NPN*	*PNP*
General-purpose, plastic-case, 350 mW, 40 V	2N4400	2N4402
Medium-low power, TO-39 case, 5 W, 60 V	2N3053	2N4036
Medium-high power, TO-220 case, 40 W, 70 V	2N6292	2N6107
High power, TO-3 case, 115 W, 60 V	2N3055	2N5875
High voltage, TO-39 case, 5 W, 250 V	2N3440	2N5416
VHF, medium-power NPN, TO-39 case, 5 W, 30 V	2N3866	
VHF, higher-power NPN, stud mounting, 11W, 40 V	2N3375	

(Power ratings for $T_{case} = 25°C$; voltages are BV_{CEO})

TABLE 2-14 (Cont.)

Other devices

N-channel junction FET, plastic, 350 mW, 25 V	2N3819
Unijunction transistor, plastic, $I_{p(max)} = 2\mu A$	2N4948
Low-power SCR, plastic, 0.8 A, 200 V	2N5064
Higher-power SCR, TO-220, 16 A, 600 V	2N6404
Triac, flat case, 4 A, 400 V, $I_{GT} = 10$ mA	2N6073 A or B
Signal diode, 20 mA, 75 V	1N914 A or B
Rectifier diode, plastic, 1 A, 400 V	1N4004
High-voltage rectifier, plastic, 1 A, 1000 V	1N4007
High-current rectifier, stud mounting, 35 A, 100 V	1N1184
Fast-switching rectifier, plastic, 1 A, 200 V	1N4935

Conventional current

Cathode   Anode
(K)

Electron flow e^-

SYMBOL

K   A

1st 2nd 3rd   Suffix
digits   letter

COLOR-BANDED

Color	Number	Suffix
Black	0	none
Brown	1	A
Red	2	B
Orange	3	C
Yellow	4	D
Green	5	E
Blue	6	F
Violet	7	G
Gray	8	H
White	9	J

1N – prefix assumed

K   A   K   A

K   A   K   A

K   A

Plastic cases

FIGURE 2-3

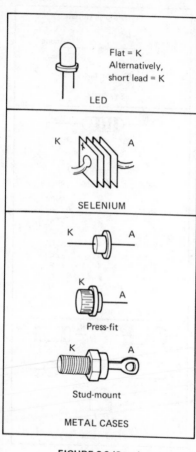

LED
Flat = K
Alternatively,
short lead = K

SELENIUM
K A

METAL CASES
K A

K A
Press-fit

K A
Stud-mount

FIGURE 2-3 (Cont.)

TO 3

$\dfrac{1.197}{1.177}$

C

E

$\dfrac{0.655}{0.675}$

B

$\dfrac{0.043}{0.038}$

$\dfrac{0.440}{0.420}$

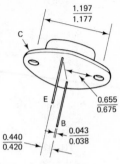

TO 18

$\dfrac{0.230}{0.209}$

C B E

$\dfrac{0.021}{0.016}$

TO 66

$\dfrac{0.958}{0.962}$

C

E

$\dfrac{0.570}{0.590}$

B

$\dfrac{0.028}{0.034}$

$\dfrac{0.190}{0.210}$

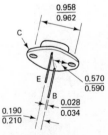

FIGURE 2-4

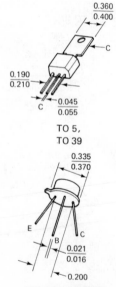

TO 92

$\dfrac{0.175}{0.205}$

E B C

$\dfrac{0.095}{0.105}$ $\dfrac{0.016}{0.019}$

TO 202

$\dfrac{0.360}{0.400}$

C

$\dfrac{0.190}{0.210}$

C $\dfrac{0.045}{0.055}$

TO 5,
TO 39

$\dfrac{0.335}{0.370}$

E   B   C

$\dfrac{0.021}{0.016}$

0.200

FIGURE 2-4 (Cont.)

90

TO 126

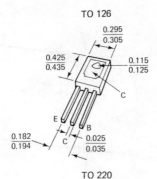

$$\frac{0.295}{0.305}$$

$$\frac{0.425}{0.435}$$

$$\frac{0.115}{0.125}$$

C

E   B

$$\frac{0.182}{0.194}$$

C $$\frac{0.025}{0.035}$$

TO 220

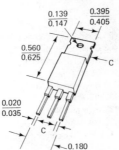

$$\frac{0.139}{0.147}$$ $$\frac{0.395}{0.405}$$

$$\frac{0.560}{0.625}$$

C

$$\frac{0.020}{0.035}$$

C

$$\frac{0.180}{0.220}$$

Most-common pinouts given
variations do occur

FIGURE 2-4 (Cont.)

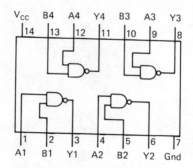

7400 Quad 2-input NAND
7403 Open-collector NANDs
7408 Quad 2-input AND
7409 Open-collector ANDs
7426 Hi-voltage NAND
7432 Quad 2-input OR
7437 Inverting NAND buffers
7438 Inv. NAND buf, O.C.
74132 NAND Schmitt triggers

FIGURE 2-5

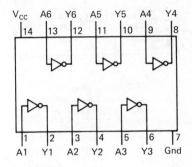

7404 Hex inverter
7405 Hex inv; open collector
7406 Hex inv. buf; HV, O.C.
7407 Hex noninv. buf; HV, O.C.
7414 Hex inv. Schmitt trigger
7416 Hex inv. buf; HV, O.C.
7417 Hex noninv. buf; HV, O.C.

FIGURE 2-5 (Cont.)

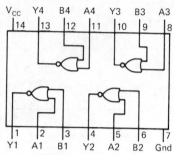

7402 Quad 2-input NOR
7401 Quad 2-input NAND
7428 Quad NOR buffers
7433 NOR buffer, O.C.
74128 NOR line drivers

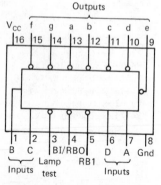

7446 BCD → 7-segment, O.C. 30-V
7447 BCD → 7-segment, O.C. 15-V
7448 BCD → 7-segment, active

FIGURE 2-5 (Cont.)

94

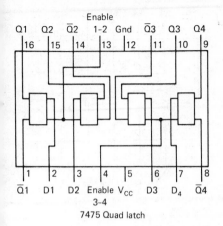

Enable

Q1	Q2	$\overline{Q2}$	1-2	Gnd	$\overline{Q3}$	Q3	Q4
16	15	14	13	12	11	10	9

1	2	3	4	5	6	7	8
$\overline{Q1}$	D1	D2	Enable	V_{CC}	D3	D_4	$\overline{Q4}$

3-4

7475 Quad latch

K1	Q1	$\overline{Q1}$	Gnd	K2	Q2	$\overline{Q2}$	J2
16	15	14	13	12	11	10	9

1	2	3	4	5	6	7	8
CLK1	PR1	CLR1	J1	V_{CC}	CLK2	PR2	CLR2

7476 Dual J–K flip-flops

FIGURE 2-5 (Cont.)

95

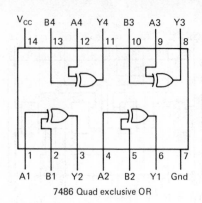

7486 Quad exclusive OR

7490 Decade counter
7493 Binary counter

FIGURE 2-5 (Cont.)

96

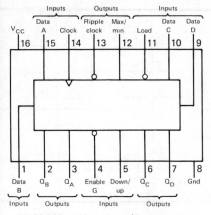

74190 BCD synchronous up/down counter
74191 Binary synchronous up/down counter

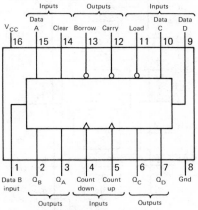

74192 BCD up/down counter
74193 Binary up/down counter
FIGURE 2-5 (Cont.)

97

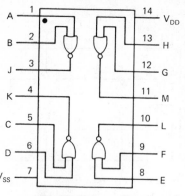

CD4001 CMOS quad NOR

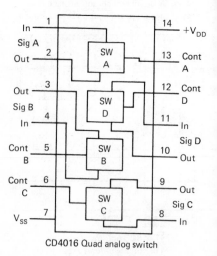

CD4016 Quad analog switch

FIGURE 2-5 (Cont.)

98

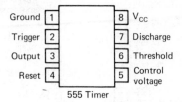

555 Timer

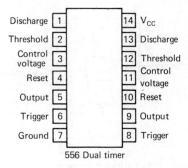

556 Dual timer

FIGURE 2-5 (Cont.)

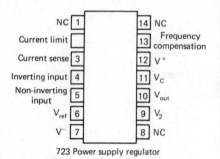

723 Power supply regulator

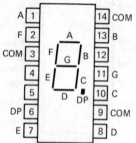

Most-common 7-segment
display pinout

FIGURE 2-5 (Cont.)

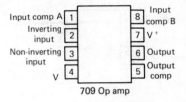

709 Op amp

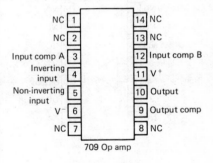

709 Op amp

FIGURE 2-5 (Cont.)

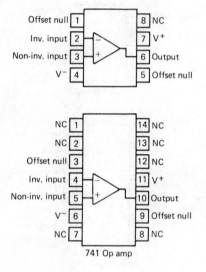

741 Op amp

FIGURE 2-5 (Cont.)

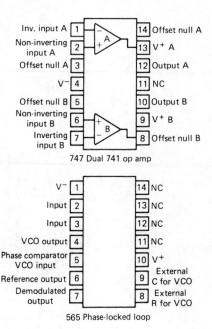

747 Dual 741 op amp

Inv. input A	1	14	Offset null A
Non-inverting input A	2	13	V⁺ A
Offset null A	3	12	Output A
V⁻	4	11	NC
Offset null B	5	10	Output B
Non-inverting input B	6	9	V⁺ B
Inverting input B	7	8	Offset null B

565 Phase-locked loop

V⁻	1	14	NC
Input	2	13	NC
Input	3	12	NC
VCO output	4	11	NC
Phase comparator VCO input	5	10	V⁺
Reference output	6	9	External C for VCO
Demodulated output	7	8	External R for VCO

FIGURE 2-5 (Cont.)

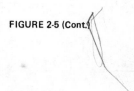

103

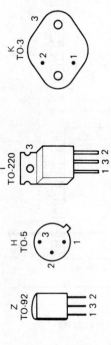

Three-terminal regulators

Positive: 1 = In, 2 = Out, 3 = Gnd
Types 109–309, 123, 130–330,
140–340, 341, 342, 78XX

Negative: 1 = Gnd, 2 = Out, 3 = In
Types 120–320, 145–345, 79XX

FIGURE 2-5 (Cont.)

2.5 MISCELLANEOUS COMPONENTS

Pilot lamps (see Table 2-15)

TABLE 2-15. Pilot-Lamp Data

Lamp	Volts	mA	Base
PR-2	2.4	500	Flange
PR-3	3.6	500	Flange
PR-4	2.3	270	Flange
PR-6	2.5	300	Flange
PR-12	5.95	500	Flange
PR-13	4.75	500	Flange
12	6.3	150	2 pin
13	3.8	300	Screw
14	2.5	300	Screw
40	6.3	150	Screw
43	2.5	500	Bayonet
44	6.3	250	Bayonet
45	3.2	350	Bayonet
46	6.3	250	Screw
47	6.3	150	Bayonet
48	2.0	60	Screw
49	2.0	60	Bayonet
50	6.3	200	Screw
51	6.3	200	Bayonet
53	14.4	120	Bayonet
55	6.3	400	Bayonet
57	14.0	240	Bayonet
112	1.2	220	Screw
222	2.2	250	Screw
233	2.3	270	Screw
239	6.3	360	Bayonet
313	28.0	170	Bayonet
1490	3.2	160	Bayonet
1819	28	40	Min. bay.
1847	6.3	150	Min. bay.
1891	14.0	240	Min. bay.

TABLE 2-15 (Cont.)

Lamp	Volts	mA	Base
1892	14.4	120	Min. bay.
NE-2(A)	125	3	Wire leads
NE-2D	125	0.7	Flange
NE-2E	125	0.7	Wire leads
NE-2H	125	1.9	Wire leads
NE-2J	125	1.9	Flange
NE-21	125	2	Bayonet
NE-30	125	12	Screw
NE-32	125	12	2-contact bay.
NE-45	125	2	Cand. screw
NE-48	125	2	2-contact bay.
NE-51	125	0.3	Min. bay.
NE-51H	125	1.2	Bayonet
NE-56	225	5	Screw
NE-57	125	2	Cand. screw
NE-58	250	2	Cand. screw
NE-83	100	5	Wire leads
NE-86	90	1.5	Wire leads

Efficiency of various light sources (light energy out/electric energy in).

Light-emitting diode	1%
Neon glow lamp	1%
Incandescent	1.5 – 3%
Mercury vapor	6%
Flourescent	7 – 12%
High–pressure sodium	16 – 22%

Batteries (see Table 2-16)

TABLE 2-16. Battery Cross Reference

Common Designation	AAA	AA	C	D	No. 6	Lantern	Transistor
Voltage (V)	1.5	1.5	1.5	1.5	1.5	6	9
Height (in.)	1.69	1.88	1.81	2.25	6.0	4.37	1.93
Width/dia. (in.)	1.39	0.53	0.94	1.25	2.50	2.63	1.03
Depth (in.)	—	—	—	—	—	2.63	0.69
NEDA number*	24	15	14	13	905	908	1604
IEC number	R-03	R-6	R-14	R-20	R-40	4R23	6F22
Eveready (alkaline)	912 (E 92)	915 (E 91)	935 (E 93)	950 (E 95)	156	509	216, 522 BP
Burgess	7	Z	1	2	6	F4M	2U6
Mallory	MN2400	MN1500	MN1400	MN1300	M905	M908	MN1604
RCA	VS074	VS034	VS035	VS036	VS0065	VS040C	VS323
Mercury type (1.4 V/cell)	—	xx9, x502	—	xx42	—	—	146x

*Suffix A indicates alkaline.

Relay contact forms (see Fig. 2-6)

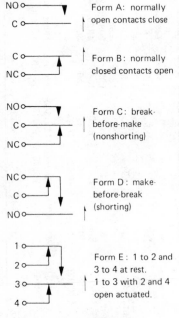

NO o———————▼
C o———————
Form A: normally open contacts close

C o———————
NC o——————▲
Form B: normally closed contacts open

NO o———————▼
C o———————
NC o——————▲
Form C: break-before-make (nonshorting)

NC o——————▲
C o———————
NO o———————
Form D: make-before-break (shorting)

1 o——————▲
2 o———————▼
3 o———————
4 o——————▲
Form E: 1 to 2 and 3 to 4 at rest. 1 to 3 with 2 and 4 open actuated.

FIGURE 2-6

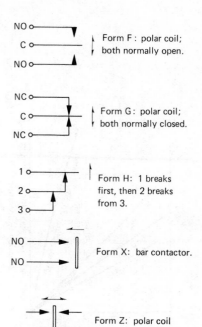

Form F : polar coil; both normally open.

Form G : polar coil; both normally closed.

Form H: 1 breaks first, then 2 breaks from 3.

Form X: bar contactor.

Form Z: polar coil

Form O: solenoid actuator only; no contacts provided

FIGURE 2-6 (Cont.)

Thermocouple characteristics (see Fig. 2-7)

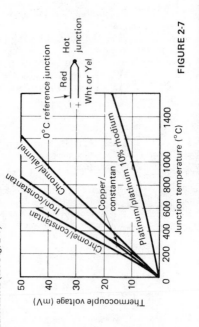

FIGURE 2-7

3

Simplified Circuit Analysis and Design

3.1 RESISTIVE CIRCUITS

Dropping resistor (Fig. 3-1): Given a voltage source V_S and a load resistance R_L with a required voltage V_{RL} (less than V_S), find a dropping resistance R_D that drops the source to the required load voltage. Disadvantage: changing load currents cause changes in load voltage.

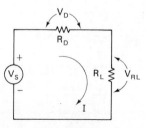

FIGURE 3-1

$$V_{RD} = V_S - V_{RL}$$

$$I = \frac{V_{RL}}{R_L}$$

$$R_D = \frac{V_{RD}}{I}$$

$$P_{RD} = IV_{RD}$$

Loaded voltage divider (Fig. 3-2): Given a voltage source V_S, a load resistance R_L with its required voltage V_O, and a regulation factor *Reg* by which V_O may rise when R_L is removed, find dividing resistors R_1 and R_2. Disadvantage: decreasing R_1 and R_2 to improve regulation causes wasteful current drain from V_S.

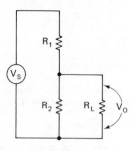

FIGURE 3-2

Analysis

$$Reg = \frac{V_{NL} - V_O}{V_O} = \frac{R_1 \parallel R_2}{R_L}$$

Design

$$R_1 = \frac{V_S}{V_O} \ \frac{Reg \, R_L}{Reg + 1}$$

$$R_2 = \frac{Reg \, R_1 R_2}{R_1 - Reg \, R_L}$$

Variable resistance network (Fig. 3-3): Determine the values of R_1, R_v, and R_2 such that a required minimum resistance R_{low} is reached when R_v is varied to zero resistance, and a required maximum R_{hi} is reached with R_v set to its maximum resistance. Note that potentiometers are commonly available only in values following a 1-2-5 sequence, and R_v must be limited to this series.

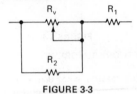

FIGURE 3-3

$$R_1 = R_{\text{low}}$$

Let R_v = next standard value above ($R_{\text{hi}} - R_1$).

$$R_2 = \frac{R_v(R_{\text{hi}} - R_1)}{R_v + R_1 - R_{\text{hi}}}$$

Variable voltage divider, unloaded (Fig. 3-4): Given two source voltages V_1 and V_2 and a bleeder current I, find values of R_1, R_v, and R_2 that will produce the required output-voltage range V_{max} to V_{min}. V_2 is often zero (ground). Any of the voltages may be negative, in which case they are treated algebraically in the formulas. If there is a load current I_L, accuracy will be impaired by a maximum amount I_L/I_{bias} (10% maximum for $I_L = 2$ mA and $I_{\text{bias}} = 20$ mA).

$$R_1 = \frac{V_1 - V_{\text{max}}}{I}$$

$$R_v = \frac{V_{\text{max}} - V_{\text{min}}}{I}$$

$$R_2 = \frac{V_{\text{min}} - V_2}{I}$$

113

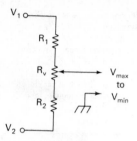

FIGURE 3-4

3.2 NETWORK THEOREMS

Thévenin's theorem states that a two-terminal network consisting of any number of resistors and voltage sources can be replaced by an equivalent circuit containing one voltage source and one resistance. The two terminals selected are usually (but not always) those across which the output voltage appears. To form the equivalent:

1. Remove a circuit element between the two terminals.

2. Determine the voltage between the two terminals with the element removed. This is V_{Th}.

3. Mentally replace all voltage sources with a short circuit and calculate the resistance of the remaining network between the two terminals. The output element is still removed. The resistance calculated is R_{Th}.

4. Replace the original network with V_{Th} and R_{Th} in series. Replace the output element and analyze the resulting simple circuit for V_{O}.

Figure 3-5 shows Thévenin's theorem used to solve a loaded time-constant problem. C is the load element removed.

$$V_{ab} = V_{Th} = V_S \frac{R_2}{R_1 + R_2}$$

$$= 12 \frac{1}{1 + 0.68} = 7.14 \text{ V}$$

$$R_{Th} = R_1 \| R_2 = 0.405 \text{ M}\Omega$$

$$V_{Cmax} = V_{Th} = 7.14 \text{ V}$$

$$\tau = R_{Th} C = 0.405 \text{ M}\Omega \times 5 \ \mu\text{F} = 2.0 \text{ s}$$

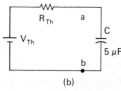

(a)

(b)

FIGURE 3-5 Thévenin's theorem

Current sources are idealizations presumed to deliver a fixed output current regardless of the load connected across them. Voltage sources are presumed to output a constant voltage regardless of load. Real voltage sources have a small resistance R_{Th} in series. Real current sources have a large resistance R_N in parallel. Figure 3-6 shows how these sources and their associated resistances can be equated.

Norton's theorem is similar to Thévenin's. It produces a current-source equivalent $I_N \| R_N$ as Thévenin produces a voltage-source equivalent.

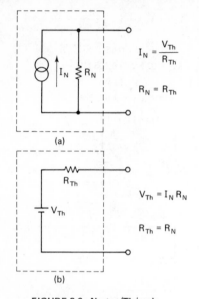

$$I_N = \frac{V_{Th}}{R_{Th}}$$

$$R_N = R_{Th}$$

(a)

$$V_{Th} = I_N R_N$$

$$R_{Th} = R_N$$

(b)

FIGURE 3-6 Norton/Thévenin

1. Remove the output element from the selected output terminals.

2. Calculate the current through a short circuit placed across these terminals. This is I_N.

3. Mentally open-circuit all current sources and short-circuit all voltage sources. Calculate R_N between the output terminals.

4. Replace the original network with $I_N \parallel R_N$.

Millman's theorem permits any number of parallel voltage sources V_n (each with its series resistance R_n) to be represented as a single voltage source V_{Th} and series resistance R_{Th}. Figure 3-7 illustrates application of the following equations:

$$R_{Th} = R_1 \parallel R_2 \parallel R_3 \cdots = \frac{1}{\dfrac{1}{R_1} + \dfrac{1}{R_2} + \dfrac{1}{R_3} + \cdots}$$

$$V_{Th} = \left(R_{Th} \frac{V_1}{R_1} + \frac{V_2}{R_2} + \frac{V_3}{R_3} + \cdots \right)$$

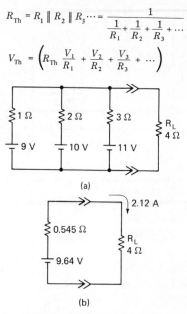

(a)

(b)

FIGURE 3-7 Millman's theorem

The superposition theorem permits solution of many networks containing multiple sources with elementary techniques. It states that the current (or voltage) response in each element of a linear bilateral network is equal to the algebraic sum of the responses produced by each source acting independently. The restriction to "linear" precludes devices such as thermistors and saturating inductors. "Bilateral" precludes one-way devices such as diodes. Capacitors, inductors, and resistors are allowable. Unused voltage sources are shorted and unused current sources are opened at each step of the analysis. Figure 3-8

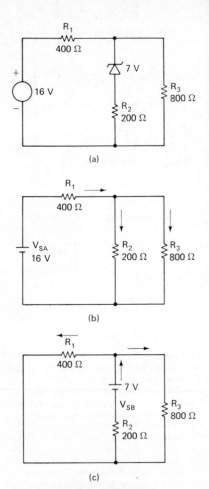

(a)

(b)

(c)

FIGURE 3-8 Superposition

118

gives an example in which a 7-V zener diode is treated as the second source. This is permissible if this diode is always in zener conduction.

$$I_{R1A} = \frac{V_{SA}}{R_1 + R_2 \| R_3} = \frac{16}{400 + 160} = 0.0286 \text{ A}$$

$$V_{R1A} = I_{R1A} R_1 = 0.0286 \times 400 = 11.43 \text{ V}$$

$$V_{R3A} = V_{SA} - V_{R1A} = 16 - 11.44 = 4.57 \text{ V}$$

$$I_{R2B} = \frac{V_{SB}}{R_2 + R_1 \| R_3} = \frac{7}{200 + 267} = 0.0150 \text{ A}$$

$$V_{R2B} = I_{R2B} R_2 = 0.0150 \times 200 = 3.00 \text{ V}$$

$$V_{R3B} = V_{SB} - V_{R2B} = 7.00 - 3.00 = 4.00 \text{ V}$$

$$V_{R3} = V_{R3A} + V_{R3B} = 4.57 + 4.00 = 8.57 \text{ V}$$

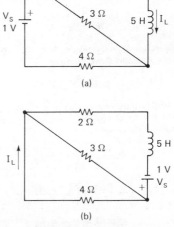

FIGURE 3-9 Reciprocity

The reciprocity theorem applies only to single-source networks. It states that the current response in any line A caused by a voltage source in any line B will be the same if the source is moved to line A and the current is measured in line B. Figure 3-9 illustrates the idea.

$$\tau = \frac{L}{R_2 + R_3 \| R_4} = \frac{5}{2 + 1.71} = 1.35 \text{ s}$$

$$I_L \text{(final)} = I_{R4} \quad \text{with } L \text{ a short circuit}$$

$$= \frac{3}{7} I_{\text{total}} = \frac{3}{7} \frac{1 \text{ V}}{2\Omega + 3\Omega \| 4\Omega} = 0.115 \text{ A}$$

Delta–wye conversion of part of a circuit will often permit a series combination that was not possible before, leading to the solution of the circuit. Figure 3-10 shows the delta and wye configurations and notation. The Δ and Y are sometimes termed π and T, respectively.

$$R_1 = \frac{R_A R_C}{R_A + R_B + R_C} \qquad R_2 = \frac{R_B R_C}{R_A + R_B + R_C}$$

$$R_3 = \frac{R_A R_B}{R_A + R_B + R_C}$$

If $R_A = R_B = R_C = R_\Delta$, then $R_1 = R_2 = R_3 = R_Y$, and the equations above reduce to

$$R_Y = \frac{R_\Delta}{3}$$

Wye–delta conversion of part of a circuit may permit new parallel combinations, leading to the solution of the circuit.

$$R_A = \frac{R_1 R_2 + R_1 R_3 + R_2 R_3}{R_2}$$

$$R_B = \frac{R_1 R_2 + R_1 R_3 + R_2 R_3}{R_1}$$

$$R_C = \frac{R_1 R_2 + R_1 R_3 + R_2 R_3}{R_3}$$

Where all Y-resistors are equal, all Δ-resistors will be equal, and

$$R_\Delta = 3R_Y$$

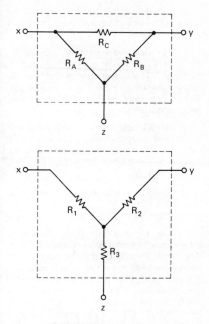

FIGURE 3-10 Delta-Wye

3.3 GENERAL CIRCUIT-ANALYSIS TECHNIQUES

Mesh analysis, also called loop analysis, is capable of solving any linear resistive circuit. Extensions of the technique permit solution of any resistive–reactive circuit. The method is based upon Kirchhoff's voltage law: *The sum of the*

121

voltage drops (resistive) is equal to the sum of the voltage rises (sources) for every closed path around a circuit.

1. Draw the circuit in a planar form (no wires crossing) if possible. Convert any current sources to voltage sources.

2. Draw clockwise current-indicating arrows within each loop, marking them I_1, I_2, and so on. Figure 3-11 shows an example. Conventional current (positive to negative) is indicated, but electron flow could be used. The only requirement is consistency.

3. Set up a table for the equations that will be obtained from the circuit. The number of rows equals the number of loops. There are three columns: A and B to the left of the equal signs and C to the right.

4. For each equation N, the column-A term is the loop current I_N times the sum of the resistances through which I_N passes.

5. The column-B term is subtracted from the column-A term. For each equation N it consists of any resistance(s) that carry another current besides I_N times that other current, I_X. It is possible that this mutual resistor may carry more than one other current besides I_N, in which case the column-B term will have the form $-R_4(I_5 + I_6)$. It is also possible that loop N will have two or more resistors carrying current from two or more other loops, in which case there will be two or more column-B terms: $-R_7I_8 - R_9I_{10}$.

6. The column-C term is the algebraic sum of the voltage sources through which I_N passes. A source is positive if it is acting in the direction indicated by I_N and negative if it acts against I_N.

122

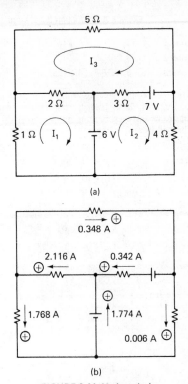

(a)

(b)

FIGURE 3-11 Mesh analysis

7. Solve the resulting simultaneous equations for each I_N by substitution or by *determinants* (see Section 1-16). Negative values for I simply mean that the current for that loop is actually counterclockwise. Figure 3-11(a) and (b) show presumed loop currents and actual branch currents, respectively.

	Column A	Column B	Column C
Loop 1	$(1+2)I_1$	$-2I_3$	$= -6$
Loop 2	$(3+4)I_2$	$-3I_3$	$= 6-7$
Loop 3	$(5+3+2)I_3$	$-3I_2 - 2I_1$	$= 7$

After collecting terms we will solve equations (1) and (2) for I_1 and I_2, respectively, and substitute these expressions into equation (3). This will leave I_3 as the only unknown in equation (3).

$$(1) \qquad 3I_1 \qquad - 2I_3 = -6$$

$$(2) \qquad 7I_2 - 3I_3 = -1$$

$$(3) \qquad -2I_1 - 3I_2 + 10I_3 = 7$$

$$(1) \qquad I_1 = \frac{-6}{3} + \frac{2I_3}{3} = \frac{2}{3}I_3 - 2$$

$$(2) \qquad I_2 = -\frac{1}{7} + \frac{3I_3}{7} = \frac{3}{7}I_3 - \frac{1}{7}$$

$$(3) \qquad -2\left(\frac{2}{3}I_3 - 2\right) - 3\left(\frac{3}{7}I_3 - \frac{1}{7}\right) + 10I_3 = 7$$

$$\left(10 - \frac{4}{3} - \frac{9}{7}\right)I_3 = 7 - 4 - \frac{3}{7}$$

$$I_3 = \frac{2.57}{7.38} = 0.348 \text{ A}$$

$$(1) \qquad 3I_1 = -6 + 2(0.348)$$

$$I_1 = -1.768 \text{ A}$$

$$(2) \qquad 7I_2 = -1 + 3(0.348)$$

$$I_2 = 0.006 \text{ A}$$

Nodal analysis is based on Kirchhoff's current law: *The algebraic sum of the currents leaving and entering any node of a circuit is zero.* A *node* is defined as the junction of two or more branches. Computer circuit-analysis programs are commonly based on nodal analysis.

1. Convert all voltage sources to current sources (Fig. 3-6).

2. Choose one node (usually ground) as the reference. Label all other nodes V_1, V_2, and so on. (see Fig. 3-12).

3. Set up a table to form the node equations. There are three columns and a number of rows equal to the number of nodes (not including the reference node).

4. The column-A term is the sum of the conductances tied to node N times V_N.

5. The column-B terms are the conductances tied to node N and another node X, times V_X.

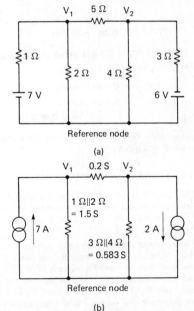

(a)

(b)

FIGURE 3-12 Node analysis

125

Node X does not include the reference node. There may be several column-B terms. Each is subtracted from the column-A term.

6. The column-C term, to the right of the equal sign, is the algebraic sum of all current sources tied to node N. A source is termed positive if it supplies current to the node and negative if it takes current away.

	Column A		Column B	Column C
Node 1	$(1.5 + 0.2)V_1$	$-$	$0.2V_2$	$= 7$
Node 2	$(0.583 + 0.2)V_2$	$-$	$0.2V_1$	$= -2$
(1)	$1.7V_1$	$-$	$0.2V_2$	$= 7$
(2)	$-0.2V_1$	$+$	$0.783V_2$	$= -2$

Multiplying (2) by 8.5 and adding the result to (1) eliminates V_1.

$$-1.7V_1 + 6.656V_2 = -17$$
$$6.456V_2 = -10$$
$$V_2 = -1.549 \text{ V}$$

Substituting this value for V_2 into (1) above yields $V_1 = 3.935$ V.

3.4 AC CIRCUIT ANALYSIS

This section is limited to circuits containing linear resistive and reactive elements and driven by sinusoidal waveforms of a single fixed frequency.

Elementary ac circuits that can be reduced to a single resistance in series or parallel with a single reactance may be solved using only Ohm's law and the resistance-reactance combination formula. (see Fig. 1-8). In the example of Fig.3-13, we determine the output across a series-tuned circuit (a) shunted by a load resistance at a frequency 1% above resonance. Generator

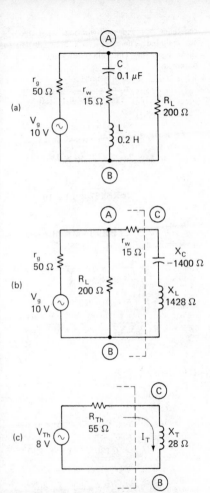

FIGURE 3-13 AC circuit simplification

127

resistance r_g and coil-winding resistance r_w are included. The circuit is redrawn in (b) to show how the Thévenin equivalent at (c) is obtained.

$$f_r = \frac{1}{2\pi\sqrt{LC}} = \frac{1}{2\pi\sqrt{0.2 \times 0.1 \times 10^{-6}}}$$
$$= 1125 \text{ Hz}$$

$$f = 1.01 f_r = 1137 \text{ Hz}$$

$$X_L = 2\pi f L = 2\pi \times 1137 \times 0.2 = 1428\,\Omega$$

$$X_C = \frac{1}{2\pi f C} = \frac{-1}{2\pi \times 1137 \times 0.1 \times 10^{-6}}$$
$$= -1400\,\Omega$$

$$X_T = X_L + X_C = 1428 - 1400 = 28\,\Omega$$

$$V_{\text{Th}} = V_G \frac{R_L}{R_L + r_g} = 10 \frac{200}{200 + 50} = 8.0\text{V}$$

$$R_{\text{Th}} = r_w + R_L \parallel r_g = 15 + 200 \parallel 50 = 55\,\Omega$$

$$Z_T = \sqrt{R_{\text{Th}}^2 + X_T^2} = \sqrt{55^2 + 28^2}$$
$$= 61.7\,\Omega$$

$$I_T = I_{XT} = I_{\text{CB}} = I_{\text{ACB}} = \frac{V_{\text{Th}}}{Z_T} = \frac{8.0}{61.7}$$
$$= 0.130 \text{ A}$$

$$Z_{\text{ACB}} = \sqrt{r_w^2 + X_T^2} = \sqrt{15^2 + 28^2} = 31.8\,\Omega$$

$$V_{\text{AB}} = I_T Z_{\text{ACB}} = 0.130 \times 31.8 = 4.12 \text{ V}$$

Series-parallel R-X conversions (see Fig. 1-9) often permit a more complex circuit to be reduced to single R and X components. The example illustrated in Fig. 3-14 demonstrates that the output of a Pi-section LC low-pass filter is down by a factor of 8 at twice the cutoff frequency (–18 dB/octave rolloff). In the passband r_g and R_L divide the 200 V, leaving $V_o' = 100$ V. At $2f_c$ we expect $V_o = 1/8 V_o' = 12.5$V. The circuit will be simplified in four stages. Voltage C-A is found in stage (d) and transferred to C-A in stage (b) where voltage division leads directly to V_{B-A}, which is V_o.

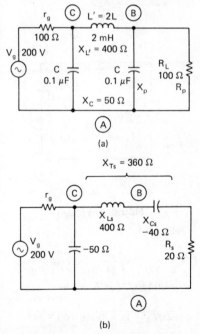

(a)

(b)

FIGURE 3-14 Parallel/series conversions

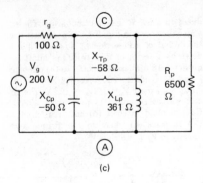

(c)

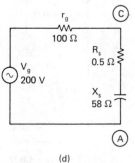

(d)

FIGURE 3-14 (Cont.)

Stage (a):

$$f_c = \frac{1}{2\pi\sqrt{LC}} = \frac{1}{2\pi\sqrt{0.001 \times 0.1 \times 10^{-6}}}$$

$$= 15\,915 \text{ Hz}; \quad 2f_c = 31\,830 \text{ Hz}$$

$$X_C \quad \frac{1}{2\pi f C} = \frac{1}{2\pi \times 31\,830 \times 0.1 \times 10^{-6}}$$

$$= 50 \ \Omega$$

$$X_L{'} = 2\pi f L' = 2\pi \times 31\,830 \times 0.002 = 400 \ \Omega$$

Stage (b):

$$R_s = X_p \frac{X_p R_p}{X_p{}^2 + R_p{}^2} = 50 \frac{50 \times 100}{50^2 + 100^2} = 20 \ \Omega$$

$$X_{Cs} = -R_p \frac{X_p R_p}{X_p{}^2 + R_p{}^2} = -100 \frac{50 \times 100}{50^2 + 100^2} = -40 \ \Omega$$

$$X_{Ts} = X_{Ls} + X_{Cs} = 400 - 40 = 360 \ \Omega$$

Stage (c):

$$R_p = \frac{X_{Ts}{}^2 + R_s{}^2}{R_s} = \frac{360^2 + 20^2}{20} = 6500 \ \Omega$$

$$X_{Lp} = \frac{X_{Ts}{}^2 + R_s{}^2}{X_{Ts}} = \frac{360^2 + 20^2}{360} = 361 \ \Omega$$

$$X_{Tp} = \frac{X_{Lp} X_{Cp}}{X_{Lp} + X_{Cp}} = \frac{361 \, (-50)}{361 - 50} = -58 \ \Omega$$

Stage (d):

$$R_s = X_{Tp} \frac{X_{Tp} R_p}{X_p{}^2 + R_p{}^2} = 58 \frac{58 \times 6500}{58^2 + 6500^2} = 0.5 \ \Omega$$

$$X_s = R_p \frac{X_{Tp} R_p}{X_{Tp}{}^2 + R_p{}^2} = 6500 \frac{58 \times 6500}{58^2 + 6500^2} = 58 \ \Omega$$

$$V_{CA} = V_g \frac{Z_{CA}}{Z_T} = 200 \frac{\sqrt{0.5^2 + 58^2}}{\sqrt{100.5^2 + 58^2}} = 100 \ V$$

Stage (b):

$$V_{BA} = V_{RL} = V_{CA} \frac{Z_{BA}}{Z_{CBA}}$$

$$= 100 \frac{\sqrt{20^2 + 40^2}}{\sqrt{20^2 + 360^2}} = 12.4 \text{ V}$$

AC analysis by complex algebra: The network theorems of Section 3.2 are valid for ac resistive–reactive circuits *provided that the quantities R, V, and I are replaced by the complex quantities. Z, V, and I.* Section 1.7 gives the form and rules for manipulation of complex quantities. It bears repeating that complex addition $(V_1 + V_2)$ is done with the quantities in rectangular form. Simply summing the magnitudes will produce incorrect results unless the phase angles are the same. Multiplication and division *(IZ)* are done in polar form. Multiplying and dividing magnitudes alone will produce results of correct magnitude, but phase-angle information, which may be needed for later additions, will be lost.

A Wien-bridge notch filter with a load resistance is analyzed at $f = \frac{2}{3} f_{notch}$ in the example of Fig. 3–15. The circuit is reduced to a single-loop Thévenin equivalent with the aid of series-parallel R–X conversion (see Fig. 1–9) and complex algebra. Every trick in the book should be used to simplify ac circuits before actually starting the solution, since complex algebra usually entails an immense amount of work. Let $R = 2 \text{k}\Omega$ and $C = 0.1 \ \mu\text{F}$. The notch frequency is

$$f_c = \frac{1}{2\pi RC} = \frac{1}{2\pi \times 2000 \times 0.1 \times 10^{-6}}$$

$$= 796 \text{ Hz}$$

$$f = \frac{2}{3} f_c = 531 \text{ Hz} \qquad X_C = 3000 \ \Omega \text{ at } f$$

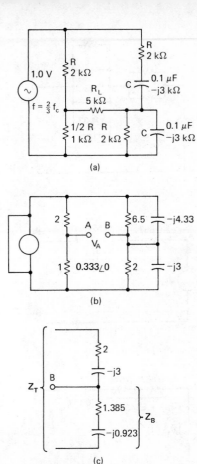

FIGURE 3-15 Wien bridge analyzed by complex algebra

133

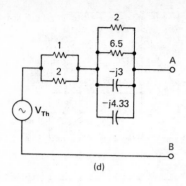

(d)

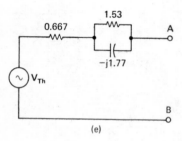

(e)

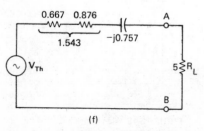

(f)

FIGURE 3-15 (Cont.)

134

The problem will be done with V in volts, R and X in kilohms, and I in milliamperes.

V_B is found from Fig. 3-15 (c):

$$Z_T = 3.385 - j3.923 = 5.182 \ \underline{/-49.2°} \ \Omega$$

$$I = \frac{V}{Z_T} = \frac{1/0}{5.182 \ \underline{/-49.2}} = 0.1930 \ \underline{/49.2°} \ A$$

$$V_B = IZ_B = 0.1930 \ \underline{/49.2°} \ \times 1.664 \ \underline{/-33.7°}$$

$$= 0.3212 \ \underline{/15.5°} = (0.3085 + j0.0858) \ V$$

$$V_{Th} = V_A - V_B = (0.3333 + j0) - (0.3095 + j0.0858)$$

$$= 0.0238 - j0.0858 = 0.0890 \ \underline{/-74.5°} \ V$$

The current and load voltage are found from Fig. 3-15 (f).

$$I = \frac{V_{Th}}{Z_T} = \frac{0.0238 - j0.0858}{6.543 - j0.757}$$

$$= \frac{0.0890 \ \underline{/-74.5°}}{6.587 \ \underline{/-6.6°}} = 0.0135 \ \underline{/-67.9°} \ A$$

$$V_{RL} = IR_L = 0.0135 \ \underline{/-67.9°} \ \times 5$$

$$= 0.0675 \ \underline{/-67.9°} \ V$$

The 5-kΩ load resistor thus drops V_{RL} to 76% of the no-load V_{Th} and shifts the phase of the output 6.6°.

Mesh and nodal analysis for ac circuits follow the rules given in Section 3.3 for dc circuits, where R, V, and I become the complex quantities Z, V, and I. The mathematics is quite tedious, and the chances of a human operator analyzing a nontrivial problem without error are slim. In a thoroughly debugged computer program, however, the method is outstanding.

The twin-tee notch filter of Fig. 3–16 will be analyzed at $f = \frac{2}{3} f_c$ as an example. Notice that the circuit does not lend itself to reduction by series-parallel R-X conversion or Thévenin's theorem.

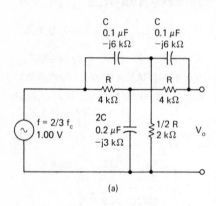

(a)

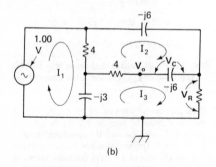

(b)

FIGURE 3-16 Twin-tee filter analyzed by mesh analysis

Calculations for Figure 3-16:

	Column A		Column B₁		Column B₂		Column C
Loop 1	$(4 - j3)\,I_1$	$-$	$(4)\,I_2$	$-$	$(-j3)\,I_3$	$=$	1
Loop 2	$(8 - j12)\,I_2$	$-$	$(4)\,I_1$	$-$	$(4 - j6)\,I_3$	$=$	0
Loop 3	$(6 - j9)\,I_3$	$-$	$(-j3)\,I_1$	$-$	$(4 - j6)\,I_2$	$=$	0
Equation 1	$(4 - j3)\,I_1$	$+$	$(-4)\,I_2$	$+$	$(j3)\,I_3$	$=$	1
Equation 2	$(-4)\,I_1$	$+$	$(8 - j12)\,I_2$	$+$	$(-4 + j6)\,I_3$	$=$	0
Equation 3	$(j3)\,I_1$	$+$	$(-4 + j6)\,I_2$	$+$	$(6 - j9)\,I_3$	$=$	0

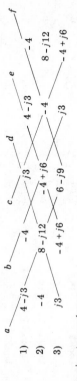

Determinant procedure:

$$D = (a_1a_2a_3) + (b_1b_2b_3) + (c_1c_2c_3) - (e_1e_2e_3) - (f_1f_2f_3)$$
$$= (-672 - j396) + (72 + j48) + (72 + j48) - (-72 + j108) - (-224 - j132) - (96 - j144)$$
$$= -328 - j132 = 353.6 \underline{/-158.1°} \ (k\Omega)^3$$

138

$$f$$

	a	b	c	d	e	f
1)	$4-j3$	-4	1		$4-j3$	-4
2)		$8-j12$	0		-4	$8-j12$
3)		$-4+j6$	0		$j3$	$-4+j6$

$$N_3 = 0 + 0 + (16 - j24) - (36 + j24) - 0 - 0 = -20 - j48 = 52.0\ \underline{/-112.6°}\ (\text{k}\Omega)^2 \cdot \text{V}$$

$$I_3 = \frac{N_3}{D} = \frac{52.0\ \underline{/-112.6°}}{353.6\ \underline{/45.5°}} = 0.147\ \underline{/-158.1°} = (0.103 + j0.105)\ \text{mA}$$

139

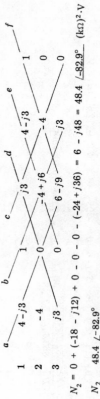

	a	b	c	d	e	f
1	4−j3	1	0	−4+j6	4−j3	1
2	−4		−4		j3	0
3		j3	6−j9		j3	0

$$N_2 = 0 + (-18 - j12) + 0 - 0 - 0 - (-24 + j36) = 6 - j48 = 48.4\ \underline{/-82.9^\circ}\ (\text{k}\Omega)^2 \cdot \text{V}$$

$$I_2 = \frac{N_2}{D} = \frac{48.4\ \underline{/-82.9^\circ}}{353.6\ \underline{/-158.1^\circ}} = 0.137\ \underline{/75.2^\circ} = (0.035 + j0.132)\ \text{mA}$$

$$V_R = I_3 R = (0.103 + j0.105)X_C = (0.206 + j0.210)\ \text{V}$$

$$V_C = (I_3 - I_2)X_C = [(0.103 + j0.105) - (0.035 + j0.132)]\,(-j6) = (-0.162 - j0.408)\ \text{V}$$

$$V_O = V_R + V_C = (0.206 + j0.210) + (-0.162 - j0.408) = 0.044 - j0.198 = (0.203\ \underline{/-77.5^\circ})\ \text{V}$$

The results show the output at $\tfrac{2}{3}f_c$ to be down from V_S by 13.9 dB and to be lagging V_S by 77.5°.

3.5 NONSINUSOIDAL EXCITATION

The square wave is the most common non-sinusoidal function, and many circuits can be reduced to single-RL or single-RC branches driven by a square wave. These can be analyzed by elementary methods.

Short-time-constant RL and RC circuits, where the time constant is much less than the switching rate ($\tau \ll t$) can be analyzed with the formulas of Section 1.4.

Long-time-constant RC circuits ($t \ll RC$) can often be analyzed with the following tools:

1. Thevenin's theorem.
2. $Q_{chg} = Q_{dischg}$ or $I_{chg} t_{chg} = I_{dischg} t_{dischg}$.
3. $C\Delta V = I\Delta t$, where ΔV is change in capacitor voltage, I is capacitor current (assumed to be reasonably constant), and Δt is time of charge or discharge ($\Delta t \ll \tau$).

The example of Fig. 3–17 depicts an 11-V 1-ms pulse with a 10 ms repetition time and 5-Ω internal resistance driving a half-wave rectifier and simple capacitor filter. We will find the v_o waveform. Note that $\tau \gg t$ for both charge and discharge.

$$I_{chg} = \frac{V_{Th} - v_o}{R_{Th}} \qquad I_{dischg} = \frac{v_o}{R_L}$$

$$I_{chg} t_{on} = I_{dischg} t_{off}$$

$$\frac{V_{Th} - v_o}{R_{Th}} t_{on} = \frac{v_o}{R_L} t_{off}$$

$$v_o = \frac{V_{th}}{1 + \frac{R_{Th} t_{off}}{R_L t_{on}}} = \frac{9.33}{1 + \frac{4.67 \times 9 \text{ ms}}{70 \times 1 \text{ ms}}} = 5.83 \text{ V}$$

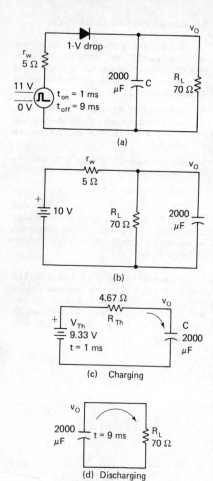

(a)

(b)

(c) Charging

(d) Discharging

FIGURE 3-17

$$\Delta v_o = \frac{I_L t_{off}}{C} = \frac{v_o t_{off}}{R_L C} = \frac{5.83 \times 9 \text{ ms}}{70 \times 2 \text{ mF}} = 0.37 \text{ V p-p}$$

Long-time-constant RL circuits ($T \gg L/R$), in which the inductor does not saturate, can be analyzed with the following formulas.

1. Thévenin's theorem.
2. $V_{chg} t_{chg} = V_{dischg} t_{dischg}$; which states that the average voltage across an inductor must be zero (plus half cycle equals minus half cycle).
3. $L\Delta I = V\Delta t$, where ΔI is the change in current through the inductor in time Δt, and V is the voltage across the inductor during Δt. V and bulk current I are assumed constant over Δt since $\tau \gg t$.

The example of Fig. 3–18 shows a pulse transformer switched on for 0.3 ms and off for 0.1 ms by a transistor. Load voltage, inductor current, and supply current are found.

$$I_{L(av)} = \frac{V_{dischg}}{R_{refl}}, \text{ but } V_{dischg} = \frac{V_{chg} t_{chg}}{t_{dischg}}, \text{ and}$$

$$V_{chg} = V_{Th} - V_{R Th} = V_{Th} - I_{L(av)} R_{Th}, \text{ so}$$

$$I_{L(av)} = \frac{(V_{Th} - I_{L(av)} R_{Th}) t_{chg}}{R_{refl} t_{dischg}},$$

which reduces to

$$I_{L(av)} = \frac{V_{Th} t_{chg}}{R_{refl} t_{dischg} + R_{Th} t_{chg}}$$

$$= \frac{9 \times 0.3}{(9 \times 0.1) + (0.9 \times 0.3)} = 2.31 \text{ A}$$

$$V_{L(chg)} = V_{Th} - I_{L(av)} R_{Th}$$

$$= 9 - (2.31 \times 0.9) = 6.92 \text{ V}$$

143

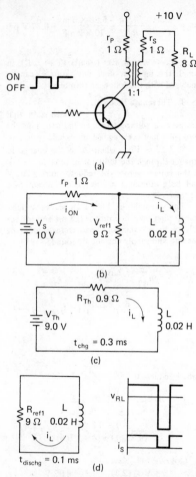

FIGURE 3-18 Long τ RL circuit

$$V_{L(\text{dischg})} = I_{L(\text{av})} R_{\text{refl}} = 2.31 \times 9 = 20.8 \text{ V}$$

$$V_{RL(\text{chg})} = V_{L(\text{chg})} \frac{R_L}{R_S + R_L} = 6.92 \times \frac{8}{9} = 6.15 \text{ V}$$

$$V_{RL(\text{dischg})} = V_{L(\text{dischg})} \frac{R_L}{R_S + R_L} = 20.8 \times \frac{8}{9} = 18.5 \text{ V}$$

$$V_{R(\text{refl})} = V_S - V_{RP}$$

$$(I_{\text{ON}} - I_L) R_{\text{refl}} = V_S - I_{\text{ON}} R_P$$

$$I_{\text{ON}} = \frac{V_S + I_L R_{\text{refl}}}{R_{\text{refl}} + R_P} = \frac{10 + (2.31 \times 9)}{9 + 1} = 3.08 \text{ A}$$

$$\Delta I = \frac{V \Delta t}{L} = \frac{6.92 \times 0.3 \text{ ms}}{0.02} = 0.10 \text{ A}$$

$$P_T = I_{\text{ON}} V_S \, Duty = 3.08 \times 10 \times \frac{3}{4} = 23.1 \text{ W}$$

$$P_{RL} = \frac{V_{RL}^2}{R_L} Duty(\text{chg}) + \frac{V_{RL}^2}{R_L} Duty(\text{dischg})$$

$$= \frac{6.15^2}{8} \times \frac{3}{4} + \frac{18.5^2}{8} \times \frac{1}{4} = 14.2 \text{ W}$$

3.6 REAL–TRANSFORMER EQUIVALENT CIRCUITS

Power transformers and audio transformers behave like the circuit of Fig. 3–19(a). Except for very large transformers, skin effect is negligible, so measured dc winding resistances may be used for r_p and r_S.

L_p and L_S behave like the ideal transformer of Section 1–8. Any voltage of whatever waveshape appearing across L_p will also appear, scaled by a factor of n, across L_S. Any current through L_S will appear, scaled by a factor of n, in L_p. Impedances across L_S are reflected across L_p by

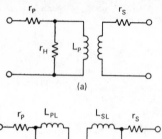

(a)

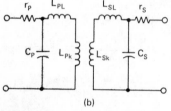

(b)

FIGURE 3-19 Real transformer equivalents

a factor of $1/n^2$. The load across L_S appears to be driven by the impedance, which drives L_P scaled by a factor of n^2.

Resistance r_H represents hysteresis loss in the core. It may be nonlinear, but is usually many times higher than X_{LP}.

Video and RF transformers behave like the circuit of Fig. 3-19(b). Skin effect (Section 1-5) will generally increase r_P and r_S over their dc values. The coils are assumed to have an air or slug-type ferrite core, so they are not perfectly coupled magnetically. The subscripts k and L represent coupled and leakage reactances, respectively.

$$L_{PL} + L_{Pk} = L_P \qquad L_{SL} + L_{Sk} = L_S$$

$$k = \frac{L_{Pk}}{L_P} = \frac{L_{Sk}}{L_S}$$

146

Measured values for coefficient of coupling k for air-core coils:

bifilar (two strands wound together)	$k \sim 0.95$
single layers, primary over secondary, $20t$	$k \sim 0.90$
end-to-end coils, length = $\frac{1}{2}$ diameter	$k \sim 0.35$
end-to-end coils, length = 2 diameters	$k \sim 0.10$
in-line, separated by one length	$k \sim 0.02$

C_p and C_S represent stray winding capacitance, which ranges from 0.01 to 0.2 pF per turn.

3.7 POWER-SUPPLY-CIRCUIT ANALYSIS

Common rectifier circuits and their output voltages are given in Fig. 3–20. Ripple frequency and typical ratios of rms secondary current to dc load current are also given. The circuits at (a), (b), and (c) can feed choke-input filters instead of the capacitors shown, in which case I_S will be about 70% of the value listed. The diode peak-inverse voltages listed are values encountered in normal service and should be multiplied by a safety factor of 2 as a minimum to cover for line transients. Capacitors should be selected with working-voltage ratings 1.5 times the operating voltages shown.

Rectifier output voltage will drop under load because of the IR drop when the first capacitor is charged by pulses of current through the transformer winding resistance. An estimate of this voltage drop can be given as

$$V_{\text{drop}} \approx 2.5 I_{s\,(\text{rms})}\, r_w$$

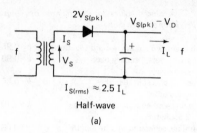

Half-wave

(a)

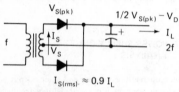

Full-wave center-tapped

(b)

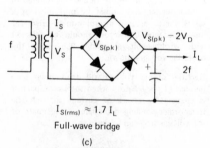

Full-wave bridge

(c)

FIGURE 3-20 Rectifiers

148

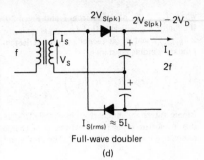

Full-wave doubler

(d)

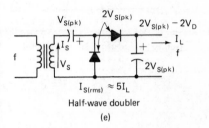

Half-wave doubler

(e)

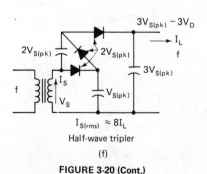

Half-wave tripler

(f)

FIGURE 3-20 (Cont.)

149

where $I_{S(rms)}$ is obtained from I_L and Fig. 3-20, the factor 2.5 indicates that $I_{S(peak)}$ is typically $2.5 I_{S(rms)}$, and r_w is secondary plus reflected-primary winding resistance.

$$r_w = R_S + R_p \left(\frac{V_S}{V_p}\right)^2$$

Ripple filters are summarized in Fig. 3-21. The first capacitor filter (a) is viewed as a charge reservoir governed by $CV = It$. The choke and

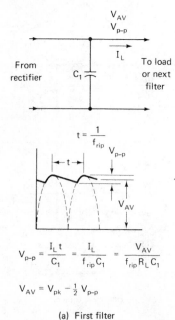

$$V_{p-p} = \frac{I_L t}{C_1} = \frac{I_L}{f_{rip} C_1} = \frac{V_{AV}}{f_{rip} R_L C_1}$$

$$V_{AV} = V_{pk} - \tfrac{1}{2} V_{p-p}$$

(a) First filter

FIGURE 3-21

150

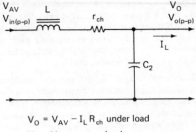

$V_O = V_{AV} - I_L R_{ch}$ under load

$V_O = V_{IN(pk)}$ at no load

$$V_{o(p-p)} \approx 0.8\, V_{in(p-p)}\, \frac{X_{C2}}{X_L} = \frac{0.02\, V_{in(p-p)}}{f_{rip}^2\, C_2\, L}$$

(b) Second filter, LC

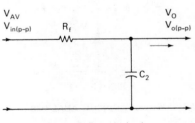

$V_O = V_{AV} - I_L R_f$ under load

$V_O = V_{IN(pk)}$ at no load

$$V_{o(p-p)} \approx 0.8\, V_{in(p-p)}\, \frac{X_{C2}}{R_f} = \frac{0.13\, V_{in(p-p)}}{f_{rip}\, C_2\, R_f}$$

(c) Second filter, RC

FIGURE 3-21 (Cont.)

151

following capacitor (b) are viewed as a voltage divider. X_L is assumed to be much greater than X_C, which is true in all practical cases. The factor 0.8 stems from the fact that the ac is a sawtooth and the fundamental sine-wave peak is about 0.8 V_{pk}. The choke is often replaced by a lighter, less costly resistor, as in (c).

Zener regulators provide load and line regulation as well as ripple reduction. Figure 3-22 gives the basic circuit. The regulation formulas neglect temperature effects on V_Z and are therefore valid only when self-heating is negligible. This is reasonably true for $I_Z \leq \frac{1}{10} I_{Z\,(max)}$.

$$R_D = \frac{V_{IN} - V_Z}{I_Z + I_L} \qquad V_{o\,(p\text{-}p)} = \frac{V_{in(p\text{-}p)} R_Z}{R_D + R_Z}$$

$$\text{Reg(load)} = \frac{R_Z}{R_L} \qquad \text{Reg(line)} = \frac{V_{IN} R_Z}{V_Z R_D}$$

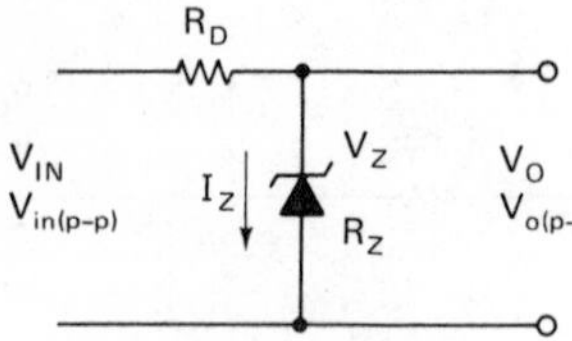

FIGURE 3-22

3.8 AMPLIFIER CIRCUIT ANALYSIS

Formulas for the analysis of transistor amplifiers can be developed from circuit-analysis techniques coupled with the following facts:

• Base current is many times smaller than emitter current (typically 40 to 300 times for small-signal transistors). Base current can thus be neglected and base bias determined by a voltage divider R_{B1}–R_{B2} if these resistors are not too large.

- The forward-biased base–emitter junction presents a dynamic (ac) resistance r_i of from $(0.025/I_E)$ to $(0.045/I_E)$ ohms. (The theoretical value determined by Shockley is $0.026/I_E$.)

- Resistance looking into the base is β times the resistance in the emitter line.

- Resistance looking into the emitter is r_i plus the resistance looking out of the base divided by β.

- Collector current is nearly equal to emitter current for both ac and dc. This allows voltage-gain formulas to be developed in a manner similiar to voltage-divider formulas: $i_c = v_c/r_c$, $i_e = v_e/r_e$, and $i_c \approx i_e$, so $v_c/v_e = r_c/r_e$, where r_c and r_e are resistances through which the collector and emitter signal currents flow.

- Resistance seen looking into the collector is very high since it is a reverse-biased diode.

- Output peak signal cannot be greater than the swing from rest (quiescent) point to saturation turn-on ($V_{CE} \to 0$).

- Output peak signal cannot be greater than the bias rest current can develop across the output load when the transistor swings to cutoff ($I_C \to 0$).

The stabilized common-emitter amplifier (Fig.3-23) provides an inverted output signal typically 5 to 100 times the voltage of the input signal. Input and output impedance levels are comparable. Input impedance is somewhat affected by transistor beta, especially in high-gain circuits, but other operating characteristics are beta-independent provided that beta is relatively high. $\beta_{min} = 40$ is a typical requirement.

DC-bias formulas:

$$V_B \approx \frac{V_{CC} R_{B2}}{R_{B1} + R_{B2}}$$

provided that $\beta (R_{E1} + R_{E2}) \gg R_{B1} \| R_{B2}$.

153

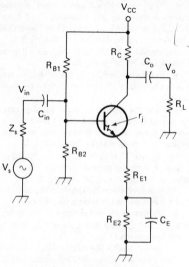

FIGURE 3-23 Stabilized CE Amp

$$V_E = V_B - V_{BE}$$

where V_{BE} typically equals 0.6 V for silicon, 0.2 V for germanium.

$$I_C \approx I_E = \frac{V_E}{R_E}$$

AC signal formulas:

$$r_j \approx \frac{0.03}{I_E} \qquad V_{o\,(\text{p-p max})} \leq 2V_{CE}$$
$$\leq 2I_C(R_C \| R_L)$$

$$A_v = \frac{V_o}{V_{in}} = \frac{R_C \| R_L}{r_j + R_{E1}}$$

154

$$Z_{in} = R_{B1} \parallel R_{B2} \parallel \beta(r_j + R_{E1})$$

$$Z_o = R_C$$

$$f_{low} \geq \frac{1}{2\pi C_{in}(Z_s + Z_{in})}$$

$$\geq \frac{1}{2\pi C_E(r_j + R_{E1})}$$

$$\geq \frac{1}{2\pi C_o(R_C + R_L)}$$

Collector self-bias, shown in Fig. 3–24, is used to improve bias stability (constant V_C). Z_{in} is

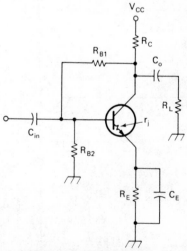

FIGURE 3-24 Collector self-bias

155

lowered because the output signal appears across R_{B1}, forcing the source to deliver more current to it. Other properties are the same as in Fig. 3-23. Note that $R_{E1} = 0$ for maximum gain in the example circuit shown.

$$V_C = \frac{R_E V_{CC} + R_C V_{BE}}{\dfrac{R_{B2} R_C}{R_{B1} + R_{B2}} + R_E}$$

$$Z_{in} = R_{B2} \parallel \frac{R_{B1}}{A_v} \parallel \beta(R_{E1} + r_j)$$

Bootstrapping, shown in Fig. 3-25, is used to

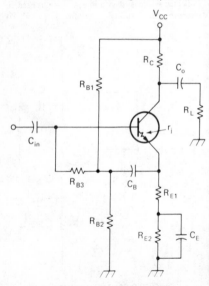

FIGURE 3-25 Bootstrapping

156

increase input impedance. It is effective only if $R_{E1} \gg r_j$, dictating much less than maximum voltage gain. Bias stability suffers since $\beta(R_{E1} + R_{E2})$ must now be much greater than $R_{B1} \| R_{B2} + R_{B3}$, and R_{B3} is large.

$$Z_{in} = R_{B3} \frac{r_j + R_{E1}}{r_j} \| \beta(r_j + R_{E1})$$

provided that

$$R_{B3} \gg R_{B1} \| R_{B2} \quad \text{and}$$
$$R_{B1} \| R_{B2} \gg R_{E1}$$

The emitter follower (Fig. 3-26) presents a high input impedance to the source and a low output impedance to the load. Bootstrapping components R_{B3} and C_B are often used to eliminate R_{B1} and R_{B2} from the input-impedance equation, as in Fig. 3-25. Voltage gain is noninverting and slightly less than 1.

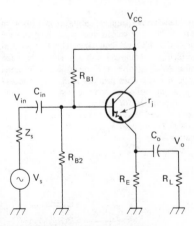

FIGURE 3-26 Emitter follower

157

$$V_E = \frac{V_{CC} R_{B2}}{R_{B1} + R_{B2}} \qquad I_E = \frac{V_E}{R_E}$$

$$A_v = \frac{R_E \| R_L}{r_j + R_E \| R_L} \qquad r_j \approx \frac{0.03}{I_E}$$

$$Z_{in} = R_{B1} \| R_{B2} \| \beta(r_j + R_E \| R_L)$$

$$Z_o = R_E \| \left(r_j + \frac{R_{B1} \| R_{B2} \| Z_s}{\beta} \right)$$

$$f_{LO} \geq \frac{1}{2\pi C_{in}(Z_s + Z_{in})}$$

$$\geq \frac{1}{2\pi C_o(Z_o + R_L)}$$

$$V_{o(\text{p-p max})} \leq 2 I_E(R_E \| R_L)$$

$$V_{o(\text{p-p max})} \leq 2 V_{CE}$$

A **transformer-coupled amplifier** is shown in Fig. 3-27. The transformer is assumed to have negligible leakage inductance (k≈1) but non-negligible winding resistance. This is true for most AF transformers. Notice that a lower turns ratio n really does produce *higher* A_v, because r_{refl} in the collector line is raised by $1/n^2$. The I_Q formula is for optimum bias point where positive and negative signal swings are equal. Lower I_Q reduces $V_{o\,(\text{max})}$. Higher I simply wastes power.

$$A_v = \frac{V_o}{V_{in}} = \frac{R_L}{n(r_j + R_E)}$$

$$I_{Q\,(\text{opt class A})} = \frac{V_{CC} - V_{CE\,(\text{sat})}}{\dfrac{r_S + R_L}{n^2} + 2(R_E + r_p)}$$

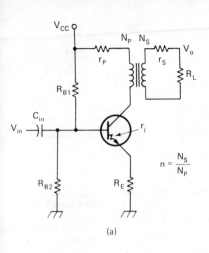

(a)

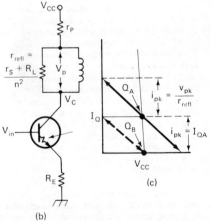

(b)

(c)

FIGURE 3-27 Transformer Coupling

159

Class A; I_Q = optimum:

$$V_{o(\text{p-p max})} = \frac{R_L (V_{CC} - V_{CE(\text{sat})})}{n(R_E + r_P) + \dfrac{r_S + R_L}{2n}}$$

Class B; I_Q = 0:

$$V_{o\ (\text{p-p max})} = \frac{R_L (V_{CC} - V_{CE(\text{sat})})}{n(R_E + r_P) + \dfrac{r_S + R_L}{n}}$$

FET amplifiers may be analyzed by formulas similar to those used for bipolar amplifiers. Here are the main differences.

• Gate-to-source voltage (instead of base current) controls device current. V_{GS} may be from -0.5 to -5 V for typical low-power N-channel junction FETs (compared to a relatively constant +0.6 V for silicon NPN transistors).

• Gate input impedance is nearly infinite [compared to $r_{b(\text{in})} = \beta r_{e(\text{line})}$]. Source-line impedance has no effect.

• Base-emitter junction resistance r_j is replaced by a fictitious source resistance $r_s = 1/y_{fs}$ Forward transfer admittance y_{fs} is sometimes called mutual conductance g_m or g_{fs}. It is the ac gain $i_{\text{out}}/v_{\text{in}}$ of the FET and is typically 3 to 15 mS for small-signal FETS.

• Minimum drain-to-source voltage $V_{DS(\text{sat})}$ is typically 2 to 4 V at medium currents, compared to 0.2 to 0.4 V for $V_{CE(\text{sat})}$.

Common-source and source-follower amplifiers are shown in Fig. 3–28. Single-supply bias is possible in (a) but it requires closely controlled FET parameters, a gate voltage divider $R_{G_1} - R_{G_2}$, and relatively high V_{DD}. Dual sup-

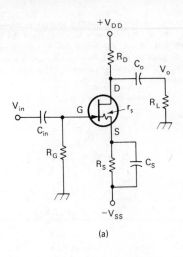

(a)

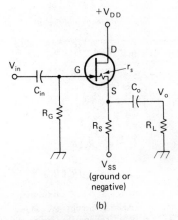

(b)

FIGURE 3-28 FET amplifiers

plies give better bias stability and are preferred. C_{in} and R_G are shown to block dc from the input, but if the source has zero dc voltage and a dc path to ground, they can be omitted. Use absolute (unsigned) values in the following equations.

Common-source amplifier (a)

$$I_D = I_S = \frac{V_{SS} + V_{GS}}{R_S}$$

$$V_D = V_{DD} - I_D R_D$$

$$Z_{in} = R_G$$

$$A_v = \frac{R_D \| R_L}{r_s} = y_{fs}(R_D \| R_L)$$

$$V_{o\ (p\text{-}p\ max)} \leq I_D(R_D \| R_L)$$
$$\leq V_{DS} - V_{DS(sat)}$$

Source-follower amplifier (b)

$$I_D = I_S = \frac{V_{SS} + V_{GS}}{R_S}$$

$$V_S = +V_{GS}$$

$$Z_{in} = R_G$$

$$A_v = \frac{R_S \| R_L}{r_s + R_S \| R_L}$$

$$V_{o\ (p\text{-}p\ max)} \leq I_D(R_S \| R_L)$$
$$\leq V_{DS} - V_{DS(sat)}$$

Operational amplifier circuits are shown in Fig. 3–29(a) (inverting) and (b) (noninverting). The inverting form is more popular because it can accept multiple inputs and because the IC input remains at virtually ground level, minimizing problems from input stray capacitance. Both circuits respond to dc as well as ac inputs.

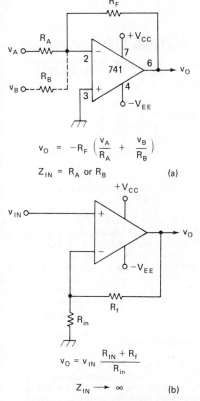

$$v_O = -R_F \left(\frac{v_A}{R_A} + \frac{v_B}{R_B} \right)$$

$$Z_{IN} = R_A \text{ or } R_B \qquad \text{(a)}$$

$$v_O = v_{IN} \frac{R_{IN} + R_f}{R_{in}}$$

$$Z_{IN} \longrightarrow \infty \qquad \text{(b)}$$

FIGURE 3-29 Inverting and noninverting op-amp

4

Units, Conversions, and Constants

4.1 THE INTERNATIONAL SYSTEM OF UNITS (SI)

Quantities are physical characteristics capable of being expressed numerically. Examples are length, pressure, and electric current. *Units* are arbitrary amounts that form the basis for measurement of quantities. Examples are miles, centimeters, pounds per square inch, and amperes. The *Système Internationale d'Unités* (SI) has been developed to provide a single, well-defined, and universally accepted unit for each quantity.

Quantity symbols:

1. Quantity symbols consist of a single letter of the English or Greek alphabet, modified by subscripts and/or superscripts where appropriate.
2. When printed, quantity symbols (and mathematical variables) appear in italic (slanted)

type. Subscripts which are quantity symbols or variables (in their own right) are also italic. Other subscripts are roman (upright). Examples: V_R, V_{max}, I_x, I_{CEO}.

3. Boldface italic type may be used to distinguish a vector quantity (I) from a scalar quantity (I). If lightface is used for vectors, then magnitudes should be distinguished by the *absolute sign* ($|I|$).

4. Because of the limited number of characters available, two quantities may be assigned the same letter symbol. To avoid confusion, an alternative letter symbol may be employed if one is listed, or the quantities may be differentiated by subscripts, or upper- and lowercase letters may be defined differently by the writer. In all cases the same quantity symbol should be retained throughout the work.

Examples: t (time), θ (temperature)

t (time), t_p (temperature)

t (time), T (temperature)

5. Several subscripts may be attached to a single quantity symbol, separated by a comma, hyphen, or parentheses if necessary for clarity. Multiple-level subscripts (subscripts attached to subscripts) are discouraged. A symbol with a superscript should be enclosed in parentheses before an exponent is added.

Electrical quantity symbols follow several additional conventions. Figure 4-1 illustrates some of them.

6. Uppercase (capital) letters are used to designate dc, rms, average (av), maximum (m) or minimum (n) values of such quantities as voltage, current and power. (The appropriate subscript may be added where clarification is

165

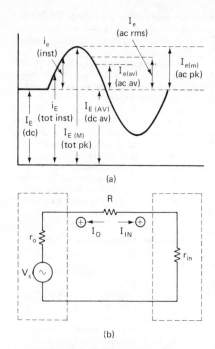

(a)

(b)

FIGURE 4-1 Quantity-symbol usage

needed.) In this case, uppercase subscripts indicate dc values, or, with the additional subscripts M or AV, maximum or average value of the total waveform. Lowercase subscripts indicate the rms value of the ac component or, with the additional subscript m, the maximum value of the ac component.

7. Lowercase (small) letters are used to designate the instantaneous value of a time-varying quantity. Here uppercase subscripts indicate instantaneous value of the total (dc + ac)

166

and lowercase subscripts indicate the instantaneous value of the ac component only.

8. Double-uppercase subscripts designate the dc supply for the element indicated. Example: V_{CC} (dc voltage supply for collector).

9. Two-letter subscripts to the voltage symbol designate voltage from the first point to the second point as a reference. Single-letter subscripts to the voltage symbol may indicate voltage across the device, or from the point designated to circuit ground as a reference.

Examples: V_{CE} (collector to emitter),
V_Z (across the zener),
V_C (collector to ground),
V_o (output to ground, ac)

10. Hyphenated subscripts may be used where two elements have the same name:

Examples: V_{B1-B2} (base 1 to base 2 of UJT)

V_{1B-2B} (base of the first to base of the second)

11. Conventional current (positive to negative) is regarded as flowing into the terminal indicated by a subscript to the current symbol. Conventional current out of the terminal gives the quantity a negative value.

12. Resistance as a circuit element is designated by uppercase R. Inherent resistance of devices such as transistors and diodes may be designated by lowercase r.

Unit symbols consist of a letter or group of letters from the English and Greek alphabets, plus a few special symbols.

167

1. Unit symbols are printed in roman (upright) type. They are never given subscripts or superscripts.

2. Lowercase letters are used except where the symbol was derived from a proper name, in which case the first letter is capitalized.

3. A space is left between the number and the unit symbol.

4. Compound units, formed by multiplication and/or division of other units, are common. Use a raised period to separate multiplication and a solidus (slash) or negative exponent to indicate division.

> Examples: N·m for newtons times meters,
> W/m^2 or $W·m^{-2}$ for watts per square meter.

5. Clarity of meaning is often served by expressing certain quantities in "phantom" units. These should be reduced to strict SI units for purposes of calculation. Examples follow.

- radian/second, cycle/second, and revolution/second reduce to 1/s.

- V/V, mV/V, and so on, for gain, regulation, common-mode rejection ratio, and so on, reduce to a unitless quantity.

- Meters/meter, $\mu in./in.$, and so on, for strain reduce to a unitless quantity.

- Ω/square for sheet resistivity reduces to Ω.

- $\Omega·cm^2/cm$ for bulk resistivity reduces to $\Omega·cm$.

- Ampere·turn for magnetomotive force reduces to A.

6. In typewritten work, substitute u for μ (adding the "tail" by hand if possible), substitute ohm for Ω, and use underline to indicate italic ($\underline{V}$ for V).

7. Unit prefixes from the list in Table 4-7 may be added in front of the SI unit symbol to avoid excessively large or small numbers and power-of-10 notation. Prefixes are generally selected to place the number in the range from 0.1 to 1000.

8. The term "billion" and the practice of separating digits into groups of three with commas should be avoided because of conflicting meanings outside the United States.

Common improper or obsolete usages of quantity and unit symbols are given in Table 4-1 with their proper forms.

TABLE 4-1. Incorrect and Correct Forms of Quantity and Unit Symbols

Not Recommended	Proper Form
$v = 15$ fps	$v = 15$ ft/s
psig $= 14.7$	$p_g = 14.7$ lb/in.^{2}
$E_p = 117$ VAC	$V_p = 117$ V ac
dB $= 12$	$\alpha = 12$ dB
$I = 1.5$ A$_{rms}$	$I_{rms} = 1.5$ A
$f = 60$ cps	$f = 60$ Hz
$f = 10$ Kc	$f = 10$ kHz
$G = 50 m\mu \mho$	$G = 50$ nS
$C = 8$ MFD	$C = 8\,\mu$F
$C = 47\,\mu\mu$F	$C = 47$ pF

169

4.2 QUANTITY, UNIT, AND UNIT-PREFIX SYMBOLS

Base units: SI is based on the independently defined units given in Table 4-2.

TABLE 4-2 BASIC SI UNITS

Quantity	Unit Name	Unit Symbol
Length, l	meter	m
Mass, m	kilogram	kg
Time, t	second	s
Electric current, I	ampere	A
Temperature, T	kelvin	K
Amount of substance	mole	mol
Luminous intensity, I_v	candela	cd
Plane angle, θ	radian	rad
Solid angle, Ω	steradian	sr

- *Kilogram* is the basic unit of mass. Kilo is not regarded here as a prefix. However, additional prefixes should be reduced. For example, microkilogram should be expressed as milligram.
- Kelvin (K), not degree kelvin or °K, is the basic unit of temperature. However the use of the word *degree* and the symbol "°" is to be continued with °C, °F, and °R.
- A radian is an angle constructed from the center of a circle such that the arc length equals the radius. There are 2π radians in a circle. A steradian is an angle constructed in a sphere such that the surface area equals the radius squared. There are 4π steradians in a sphere.

Derived units that have been given special names are listed in Table 4-3. All SI measurements are to be expressed by combinations of base and derived units.

TABLE 4-3. SI Derived Units

Quantity	Unit Name	Unit Symbol	Formula	SI Base Units
Force, F	newton	N	$kg \cdot m/s^2$	$kg \cdot m/s^2$
Pressure, stress, p	pascal	Pa	N/m^2	$kg/m \cdot s^2$
Energy, work, heat, W	joule	J	$N \cdot m$	$kg \cdot m^2/s^2$
Power, P	watt	W	J/s	$kg \cdot m^2/s^3$
Charge, Q	coulomb	C	$A \cdot s$	$A \cdot s$
Electromotive Force, V	volt	V	W/A	$kg \cdot m^2/A \cdot s^3$
Resistance, R	ohm	Ω	V/A	$kg \cdot m^2/A^2 \cdot s^3$
Conductance, G	siemens	S	A/V	$A^2 \cdot s^3/kg \cdot m^2$
Capacitance, C	farad	F	C/V	$kg \cdot m^2/s^2$
Inductance, L	henry	H	Wb/A	$kg \cdot m^2/A^2 \cdot s^2$
Magnetic flux, Φ	weber	Wb	$V \cdot s$	$kg \cdot m^2/A \cdot s^2$
Flux density, B	tesla	T	Wb/m^2	$kg/A \cdot s^2$
Frequency, f	hertz	Hz	$1/s$	s^{-1}
Luminous flux, Φ_v	lumen	lm	$cd \cdot sr$	$cd \cdot sr$
Illumination, E_v	lux	lx	lm/m^2	$cd \cdot sr/m^2$

Magnetic "hand" rules.

1. Field about a wire: Grasp a wire with the right hand, the thumb pointing in the direction of conventional current (positive to negative). The fingers curl around the wire in the direction of the magnetic lines of force (*N* to *S*).

2. Solenoid magnetic polarity: Grasp the coil with the right hand, the fingers curling around in the direction of conventional current. The thumb points to the north pole of the solenoid.

3. Direction of induced current: Point index finger of the right hand in direction of lines of force (*N* to *S*), thumb in direction of motion of conductor. Middle finger points direction of induced conventional current (positive end of wire).

4. Force on a moving charge: Point index finger of *left* hand in the direction of lines of force (*N* to *S*), middle finger in direction of conventional current (+ to -). Thumb points direction of force.

Magnetic quantities and units are less widely understood than their electrical counterparts, perhaps because magnetic measurement instruments are less common. Table 4-4 lists comparable units as an aid to conceptualization.

Light is normally measured in the *photometric* system, which includes only the portion of the spectrum visible to the human eye. The *radiometric* system includes all wavelengths of electromagnetic radiation. The two systems are compared in Table 4-5.

TABLE 4-4. Electrical and Magnetic Quantity and SI Unit Symbols

Electrical				Magnetic			
Quantity	Symbol	SI Unit		Quantity	Symbol	SI Unit	Comments
Electromotive force	V	V		Magnetomotive force	F_m	A	ampere-turns
Field strength	E	V/m		Magnetization	H	A/m	
Electric current	I	A		Magnetic flux	Φ	Wb	Weber = V·s
Current density	J	A/m^2		Flux density	B	T	tesla = Wb/m^2
Resistance	R	Ω		Reluctance	R_m	H^{-1}	$R_m = F_m/\Phi$
Resistivity	ρ	$\Omega \cdot$m		Reluctivity	ν	m/H	
Conductance	G	S		Permeance	P_m	H	
Conductivity	γ	S/m		Permeability	μ	H/m	$\mu = B/H$
Relative conductivity	μ_γ	—		Relative permeability	μ_γ	numeric	μ_0 = initial

173

TABLE 4-5. Photometric and Radiometric Quantity and SI Unit Symbols

Quantity	Photometric			Radiometric	
	Symbol	SI Unit	Comment	Symbol	SI unit
Intensity of source	I_v	cd	cd = lm/sr	I_e	W/sr
Luminous flux (power)	Φ_v	lm	lm = cd · sr	Φ_e	W
Illumination (irradiance)	E_v	lx	lx = lm/m²	E_e	W/m²
Efficacy	K	lm/W	$\dfrac{\text{visible light}}{\text{total power}}$	—	—

SI quantity and unit symbols (other than magnetic and light) are given in Table 4-6. Reserve symbols in parentheses are to be used only to avoid conflicts.

TABLE 4-6. Additional Quantity and SI Unit Symbols

Quantity	Symbol	Unit
Spatial		
Plane angle	$\theta, \phi, \alpha, \beta$	rad
Solid angle	$\Omega \ (\omega)$	sr
Length	l	m
Path length	s	m
Thickness	d, δ	m
Radius	r	m
Diameter	d	m
Area	$A \ (S)$	m^2
Volume	V, v	m^3
Time	t	s
Velocity	v	m/s
Angular velocity	ω	rad/s
Rotational speed	n	r/s
Acceleration	a	m/s^2
(of free fall)	g	m/s^2
Angular acceleration	α	rad/s^2
Mechanical		
Force	F	N
Weight	W	N
Mass	m	kg
Density	ϱ	kg/m^3
Pressure	p	Pa
Momentum	p	kg·m/s
Torque	$T \ (M)$	N·m
Rotational inertia	I, J	kg·m²
Work	W	J
Energy	E, W	J
Efficiency	η	Numerical

TABLE 4-6 (cont.)

Quantity	Symbol	Unit
Stress	σ	N/m^2
Strain	ϵ	Numerical
Thermal		
Temperature, Celsius	$t\,(\theta)$	°C
Temperature, absolute	$T\,(\Theta)$	K
Heat energy	Q	J
Heat flow rate	$\Phi\,(q)$	W
Thermal resistance	R_θ	K/W
Thermal resistivity	ϱ_θ	m·K/W
Heat capacity	C_θ	J/K
Specific Heat	c	J/K·kg
Electrical		
Charge	Q	C
Field strength	$E\,(K)$	V/m
Electromotive force	V, E	V
Current	I	A
Current density	$J\,(s)$	A/m^2
Resistance	R	Ω
Resistivity, volume	ϱ	Ω·m
Conductance, 1/R	G	S
Conductivity	γ, σ	S/m
Reactance	X	Ω
Susceptance, -1/X	B	S
Impedance	Z	Ω
Admittance, 1/Z	Y	S
Characteristic impedance	Z_0	Ω
Transadmittance	y_{ij}	S
Mutual conductance	g_m	S
Amplification factor	μ	Numerical
Quality factor, X_s/R_s	Q	Numerical

TABLE 4-6 (cont.)

Quantity	Symbol	Unit
Dissipation factor, $1/Q$	D	Numerical
Phase angle	θ, ϕ	rad
Power	P	W
Reactive "power"	$Q\,(P_q)$	var
Apparent "power"	$S\,(P_s)$	V·A
Power factor	$\cos\phi\,(F_p)$	Numerical
Period	T	s
Time constant	$\tau\,(T)$	s
Frequency	$f\,(\nu)$	Hz
Angular frequency	ω	rad/s
Resonant frequency	f_r	Hz
Critical frequency	f_c	Hz
Wavelength	λ	m
Rise time (10%–90%)	t_r	s
Fall time (90%–10%)	t_f	s
Duty factor	D	Numerical
Duration of signal element	τ	s
Signaling speed (baud)	$1/\tau$	Bd
Bandwidth	B	Hz
Noise figure	F	Numerical
Amplification (current or voltage)	A	Numerical
Gain (power)	G	Numerical
Feedback ratio	β	Numerical
Attenuation	α	Numerical

Unit prefixes are given in Table 4-7. Centi, deci, deka and hecto are to be avoided where possible. Pronounce giga as in *jig*, kilo as in *kill*, nano as in *Nancy*, pico as in *peek*, and peta as in *pet*.

177

TABLE 4-7. SI Unit Prefixes

Factor	Name	Symbol	Factor	Name	Symbol
10^1	deka	da	10^{-1}	deci	d
10^2	hecto	h	10^{-2}	centi	c
10^3	kilo	k	10^{-3}	milli	m
10^6	mega	M	10^{-6}	micro	μ
10^9	giga	G	10^{-9}	nano	n
10^{12}	tera	T	10^{-12}	pico	p
10^{15}	peta	P	10^{-15}	femto	f
10^{18}	exa	E	10^{-18}	atto	a

4.3 UNIT CONVERSIONS

Table 4-8 gives conversions from various units to the appropriate SI units.

• To convert from the SI unit to the left-column unit, *divide* by the factor given.

• To convert from one non-SI unit to another, multiply by the factor given with the known unit and divide by the factor given with the unknown unit (unit $A \rightarrow$ SI $\rightarrow$ unit B).

• To convert compound units not listed, convert each unit separately. All units except one will have the numerical coefficient 1. As an example, we convert 23 kilometers per liter to fuel consumption in miles per gallon:

$$\frac{23 \text{ km}}{1 \text{ L}} = \frac{23 \times 1000}{1 \times 10^{-3}}$$
$$\text{(given units)} \quad \text{(SI units)}$$

$$= \frac{(2300 \div 1609) \text{ mi}}{(1 \times 10^{-3} \div 3.785 \times 10^{-3}) \text{ gal}} = 54.1 \text{ mi/gal}$$
$$\text{(convert to desired units)}$$

TABLE 4-8 Unit Conversion

To Convert from	to	Multiply by
Acceleration		
ft/s^2	m/s^2	**0.3048**
in./s^2	m/s^2	**0.0254**
earth gravity, g	m/s^2	9.807
Angle		
degree	rad	0.01745
gradient	rad	0.01571
minute	rad	2.909×10^{-4}
second	rad	4.848×10^{-6}
Area		
acre	m^2	4047
barn	m^2	10^{-28}
circular mil	m^2	5.067×10^{-10}
ft^2	m^2	0.09290
hectare	m^2	10000
in.^{2}	m^2	6.454×10^{-4}
mi^2	m^2	2.590×10^{6}
yd^2	m^2	0.8361
Density (mass per unit volume)		
g/cm^3	kg/m^3	**1000**
lb/ft^3	kg/m^3	16.02
lb/in.^{3}	kg/m^3	2.768×10^{4}
lb/gal	kg/m^3	119.8
Electrical and magnetic		
decibels	nepers	0.1151
faraday	C	9.652×10^{4}
gauss	T	10^{-4}
gilbert	A·turns	0.7958
maxwell (line)	Wb	10^{-8}
oersted	A/m	79.58
Ω·cm	Ω·m	**0.01**
Ω·cmil/ft	Ω·m	1.662×10^{-9}
Ω·mm^2/m	Ω·m	10^{-6}
unit pole	Wb	1.257×10^{-7}
Energy		
British thermal unit	J	1056
calorie	J	4.190
electron volt	J	1.602×10^{-19}
erg	J	10^{-7}

TABLE 4–8 (cont.)

To Convert from	to	Multiply by
ft·lb	J	1.356
ft·poundal	J	0.04214
kilocalorie	J	4190
kW·h	J	3.6×10^6
W·s	J	1
Force		
dyne	N	10^{-5}
kg-force	N	9.807
ounce-force	N	0.2780
pound-force	N	4.448
poundal	N	0.1382
ton-force	N	8.896
Length		
angstrom	m	10^{-10}
astronomical unit	m	1.496×10^{11}
chain (surveyor's)	m	20.12
fathom	m	1.829
foot	m	0.3048
inch	m	0.0254
league	m	4828
light year	m	9.461×10^{15}
microinch	m	2.54×10^{-8}
micron	m	10^{-6}
mil	m	2.54×10^{-5}
mile (nautical)	m	1852
mile (statute)	m	1609
parsec	m	3.086×10^{16}
pica (printer's)	m	4.218×10^{-3}
point (printer's)	m	3.515×10^{-4}
rod	m	5.029
yard	m	0.9144
Light		
candle	candela	1
candle power, spherical	lumen	12.57
ft·cd (at 2870 K)	W/m²	0.005
footcandle	lux	10.76
footlambert	cd/m²	3.426
lambert	cd/m²	3183

TABLE 4–8 (cont.)

To Convert from	to	Multiply by
lumen (at 5500 Å)	W	0.0015
Mass		
carat	kg	2×10^{-4}
grain	kg	6.480×10^{-5}
gram	kg	10^{-3}
ounce (avoirdupois)	kg	0.02835
ounce (apothecary)	kg	0.03110
pound		
(avoirdupois)	kg	0.4536
slug	kg	14.59
ton (assay)	kg	0.02917
ton (long, 2240 lb)	kg	1016
ton (metric)	kg	**1000**
ton (short, 2000 lb)	kg	907.2
tonne	kg	**1000**
Power		
Btu/h	W	0.2931
horsepower	W	746
ton (refrigeration		
12 000 Btu/h)	W	3517
Pressure (force per unit area)		
atmosphere	Pa	1.013×10^5
bar	Pa	10^5
cm of mercury	Pa	1333
dyne/cm^2	Pa	0.1
gram-force/cm	Pa	98.07
inch of mercury	Pa	3386
lb-force/in.^{2} (psi)	Pa	6895
mm Hg (torr)	Pa	133.3
Temperature, heat		
Btu·in./h·ft^2·°F	W/m·K	0.1442
(thermal conductivity)		
Btu/lb·°F	J/kg·K	4187
(heat capacity)		

$$°C \rightarrow K \qquad t_K = t_{°C} + 273.15°$$

$$°F \rightarrow °C \qquad t_{°C} = (t_{°F} - 32°)/1.8$$

$$°F \rightarrow K \qquad t_K = (t_{°F} + 459.7°)/1.8$$

TABLE 4-8 (cont.)

To Convert from	to	Multiply by
°R→K	$t_K = t_{°R}/1.8$	
Torque		
dyne·cm	N·m	**10^{-7}**
kg force·m	N·m	9.807
oz force·in.	N·m	7.062×10^{-3}
lb force ·in.	N·m	0.1130
lb force·ft	N·m	1.356
Velocity		
ft/min	m/s	8.467×10^{-5}
ft/s	m/s	**0.3048**
in./s	m/s	**0.0254**
km/h	m/s	0.2778
knot	m/s	0.5144
mi/h	m/s	**0.447**
mi/s	m/s	1609
Volume		
acre-foot	m³	1233
barrel (42 gal)	m³	0.1590
board foot	m³	2.360×10^{-3}
bushel	m³	0.03524
cord (wood)	m³	3.625
cup	m³	2.366×10^{-4}
fluid ounce	m³	2.957×10^{-5}
ft³	m³	0.02832
gallon, liquid	m³	3.785×10^{-3}
in.³	m³	1.639×10^{-5}
liter (L)	m³	**10^{-3}**
ounce, fluid	m³	2.957×10^{-5}
peck	m³	8.810×10^{-3}
pint liquid	m³	4.732×10^{-4}
quart, dry	m³	1.101×10^{-3}
quart, liquid	m³	9.464×10^{-4}
tablespoon	m³	1.479×10^{-5}
teaspoon	m³	4.929×10^{-6}
yd³	m³	0.7646

NOTE: Bold-face numbers are exact. Others are given to four significant digits.

4.4 PHYSICAL CONSTANTS AND MATERIAL PROPERTIES

Selected physical constants:

Length of a 1-s pendulum at sea level
$$= 0.994\,\text{m}$$

Solar radiation intensity at earth distance
$$= 1.35\,\text{kW/m}^2$$

Mean radius of earth = 6370km

Mass of earth = 5.983×10^{24} kg

Earth gravity (g) = 9.81 m/s^2

Universal gravitational constant (G)
$$= 6.670 \times 10^{-11}\,\text{N·m}^2/\text{kg}^2$$

Standard atmospheric pressure
$$= 1.013 \times 10^5\,\text{N/m}^2$$

Boltzmann's constant (k)
$$= 1.381 \times 10^{-23}\,\text{J/K}$$

Planck's constant (h) = 6.626×10^{-34} J·s

Avogadro's number (N_A)
$$= 6.023 \times 10^{23}\,\text{mol}^{-1}$$

Permeability of vacuum (μ_0 or Γ_m)
$$= 4\pi \times 10^{-7}\,\text{H/m}$$

Permittivity of vacuum (ϵ or Γ_e)
$$= 8.854 \times 10^{-12}\,\text{F/m}$$

Velocity of sound in air at standard pressure and 0°C = 331.4m/s

Velocity of electromagnetic waves (c_0)
$$= 2.998 \times 10^8\,\text{m/s}$$

Velocity of sound in water = 1420 m/s

Velocity of sound in steel = 5103 m/s

Electron charge (e) = 1.602×10^{-19} C

Charge-to-mass ratio of the electron
$$= 1.759 \times 10^{11}\,\text{C/kg}$$

Atomic mass unit (carbon 12 = 12 amu)
$$= 1.6605 \times 10^{-27}\,\text{kg}$$

Electron mass = 5.486×10^{-4} amu

Proton mass = 1.00728 amu

Neutron mass = 1.00867 amu

Alpha particle = 4.00151 amu

Diameter of hydrogen atom = 1.06×10^{-10} m

Loudness of sound: Sound intensity is measured in decibels with 10^{-12} W/m^2 defined as zero dB. The lower limit of audibility varies with individuals but is normally within a few dB of 0 dB. The ear is most sensitive at about 3 kHz. Sensitivity at 300 Hz is down by about 20 dB.

15 dB whisper	70 dB truck engine
30 dB office background	85 dB jackhammer
55 dB average conversation	110 dB rock band

Air pressure vs. elevation:

Elevation (km)						
0	0.5	1.0	2.0	3.0	5.0	10
1.00	0.94	0.89	0.78	0.69	0.54	0.26
Pressure (atm)						

Coefficients of friction, $F_{parallel}/F_{normal}$ (numerical):

Rubber on dry pavement	0.75
Rubber on wet pavement	0.50
Wood on hardwood floor	0.25
Steel on steel (dry)	0.20
Steel on steel (oiled)	0.06
Steel runners on ice	0.04

Density at $0°$ C and standard pressure (kg/m^3):

Gold	19300	Mercury	13600
Copper	8890	Water	1000
Steel	7830	Seawater	1025
Aluminum	2700	Ice	910
Magnesium	1741	Gasoline	690

Cork	224	Propane	2.02
Woods :		Carbon dioxide	1.96
Pine	430	Air	1.29
Oak	750	Helium	0.18
Birch	640	Hydrogen	0.09

Young's modulus Y (stress/strain) and ultimate strength σ_x :

Material	Y (N/m^2 × 10^{10})	σ_x (N/m^2 × 10^8)
Aluminum	7.0	1.4
Copper	12.5	2.4
Cast iron	9.1	2.9
Mild steel	17.2	4.1
Spring steel	—	13.8
Tungsten	35	41
Magnesium	4.2	1.9

Thermal coefficients of expansion (1/K × 10^{-6}):

Linear		Volume	
Aluminum	24	Alcohol	1220
Brass	18	Gasoline	1080
Copper	17	Water at 20°C	207
Glass	8	Mercury	182
Glass (Pyrex)	3		
Steel	11		
Magnesium	27		

Thermal conductivity (W/m·K):

Silver	415	Glass	0.78
Copper	381	Brick	0.72
Aluminum	213	Wood (fir)	0.11
Steel	50	Cork board	0.037
Lead	35	Spun-fiber	
Ice	2.2	insulation	0.036
Sand, concrete	1.8		

Melting and boiling points at standard pressure (K):

Substance	Melting	Boiling
Helium	0.94	4.3
Nitrogen	63.2	77.1
Carbon dioxide	216.5	213.2
Ammonia	198.7	239.8
Alcohol, ethyl	143.2	350.6
Water	273.2	373.2
Tin	505	2530
Lead	600	1890
Copper	1360	2570
Iron	1810	3260

Heat capacities (J/kg·K):

Water	4190	Air	1010	Steel	482
Ice	2100	Aluminum	922	Copper	390
Steam	2010	Dry earth	840	Lead	126

Heats of fusion and vaporization (J/kg × 10⁶):

Water	0.335 (fusion)	2.26	(vaporization)
Ammonia	0.452 (fusion)	1.37	(vaporization)
Alcohol	0.104 (fusion)	0.855	(vaporization)

Heating values of fuels (J/kg × 10⁶):

Hydrogen	142	Fuel oil	45
Natural gas	55	Coal (hard)	33
Gasoline	47	Wood (average)	15
Kerosine	46		

5

Standards, Symbols, and Codes

5.1 FREQUENCY STANDARDS

Musical scale (Table 5–1) Equally Tempered (Hz).

Broadcast frequencies

- *AM radio:* 535 to 1605 kHz. 107 10-kHz channels with carriers on multiples of 10 kHz. Tolerance, ±10 Hz maximum.
- *FM radio:* 88.0 to 108 MHz. 100 150-kHz channels with 25-kHz guard bands on either side. Center frequencies on odd multiples of 0.1 MHz. Tolerance, ±2 kHz.
- *VHF television:* 54 to 72 MHz (channels 2, 3, and 4), 76 to 88 MHz (channels 5 and 6), 174 to 276 MHz (channels 7 to 13); see Table 5–2.

The electromagnetic spectrum (see Fig. 5–1)

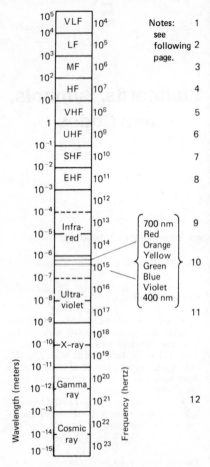

FIGURE 5-1 Electromagnetic Spectrum

1. Not allocated below 10 kHz.
   Submarine, ship-to-shore communication and navigation.

2. Reliable long-range communication with high power.
   Marine and aircraft communication and navigation.

3. Reliable medium-range communication.
   Standard AM broadcast, marine, and air communication.

4. Erratic long-distance communication via ionospheric reflection.
   International broadcast, amateur, industrial, military, marine.

5. Reliable short-range communication.
   Mobile radio, police, air traffic, TV 2-13, FM broadcast.

6. Line-of-sight communication.
   Air traffic, mobile, satellite, TV 14-70.

7. Line-of-sight communication.
   Radar, microwave telephone and television relays, satellite.

8. Radar, satellite, mobile, experimental.

9. Radiant heat.

10. Visible light.

11. Sunburn, "darklight" surveillance.

12. Nuclear radiation.

FIGURE 5-1 (Cont.)

TABLE 5-1. 88-Note Keyboard Scale

A	27.5	55.0	110.0	220.0	440.0	880.0	1760	3520
A#	29.1	58.3	116.5	233.1	466.2	932.3	1865	3729
B	30.9	61.7	123.5	246.9	493.9	987.8	1976	3951
C	32.7	65.4	130.8	261.6	523.3	1047	2093	4186
C#	34.6	69.3	138.6	277.2	554.4	1109	2217	
D	36.7	73.4	146.8	293.7	587.3	1175	2349	
D#	38.9	77.8	155.6	311.1	622.3	1245	2489	
E	41.2	82.4	164.8	329.6	659.3	1319	2637	
F	43.7	87.3	174.6	349.2	698.5	1397	2794	
F#	46.2	92.5	185.0	370.0	740.0	1480	2960	
G	49.0	98.0	196.0	392.0	784.0	1568	3136	
G#	51.9	103.8	207.7	415.3	830.6	1661	3322	

TABLE 5-2 VHF Television Frequencies

Channel TV	Lower Limit (MHz)	Channel TV	Lower Limit (MHz)
2	54	8	180
3	60	9	186
4	66	10	192
5	76	11	198
6	82	12	204
7	174	13	210

• *UHF channels 14 to 69:* Lower limit = $6 (N-14) + 470$ where N is the UHF channel number.

Picture carrier freq. = *lower limit* + 1.25 MHz
Sound carrier freq. = *lower limit* + 5.75 MHz
Tolerance = ±1kHz

• *Regional broadcasting:*

2300 – 2495 kHz
3200 – 3400 kHz
4750 – 4995 kHz

• *International broadcasting:*

5.95 – 6.20 MHz   17.70 – 17.90 MHz
9.50 – 9.775 MHz   21.45 – 21.75 MHz
11.70 – 11.975 MHz   25.60 – 26.10 MHz
15.10 – 15.45 MHz

• *Radar bands:*

L-band:	390– 1 550 MHz
S-band:	1 550– 5 200 MHz
X-band:	5 200– 10 900 MHz
K-band:	10 900– 36 000 MHz
Q-band:	36 000– 46 000 MHz
V-band:	46 000– 56 000 MHz
W-band:	56 000–100 000 MHz

- *Citizen's band (class A) frequencies (MHz):*

462.55–463.20; 464.75–464.95; 465.05–466.95

- *Amateur radio frequencies (subbands in Fig. 5-2):*

160 meters	1.800– 2.000 MHz		420– 450 MHz
80 meters	3.500– 4.000 MHz		1215–1300 MHz
40 meters	7.000– 7.300 MHz		2300–2450 MHz
Effective 1982	10.100–10.150 MHz		3300–3500 MHz
20 meters	14.000–14.350 MHz		5650–5925 MHz
Proposed	18.068–18.168 MHz		10–10.5 GHz
15 meters	21.000–21.450 MHz		24–24.25 GHz
Proposed	24.890–24.990 MHz		48–50 GHz
10 meters	28.000–29.700 MHz		71–76 GHz
6 meters	50–54 MHz		165–170 GHz
2 meters	144–148 MHz		240–250 GHz
1¼ meters	220–225 MHz		all above 300 GHz

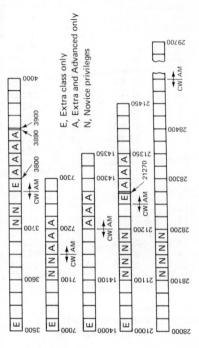

E, Extra class only
A, Extra and Advanced only
N, Novice privileges

FIGURE 5-2

- *Citizen's band (class D) (see Table 5–3)*

TABLE 5–3. Citizens' Band (Class D) Frequencies

Channel	Frequency (MHz)	Channel	Frequency (MHz)
1	26.965	21	27.215
2	26.975	22	27.225
3	26.985	23	27.255
4	27.005	24	27.235
5	27.015	25	27.245
6	27.025	26	27.265
7	27.035	27	27.275
8	27.055	28	27.285
9	27.065	29	27.295
10	27.075	30	27.305
11	27.085	31	27.315
12	27.105	32	27.325
13	27.115	33	27.335
14	27.125	34	27.345
15	27.135	35	27.355
16	27.155	36	27.365
17	27.165	37	27.375
18	27.175	38	27.385
19	27.185	39	27.395
20	27.205	40	27.405

Note: Tolerance, ±0.005%

5.2 TELEVISION AND RADIO BROADCAST STANDARDS

Television standards, U.S.A. (see also Fig. 5–3):

- *Aspect ratio:* (width/height) = 4:3.
- *Channel width:* 6 MHz.
- *Video system:* amplitude modulated; vestigial lower sideband leaving total 4.2-MHz video bandwith; picture black corresponds to

194

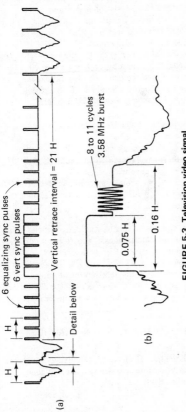

6 equalizing sync pulses

6 vert sync pulses

Vertical retrace interval = 21 H

Detail below

H

H

(a)

8 to 11 cycles
3.58 MHz burst

0.075 H

0.16 H

(b)

FIGURE 5-3 Television video signal

75% of peak carrier strength, white to 12.5% of carrier strength.

- *Sound system:* independent carrier 4.5 MHz above picture carrier, frequency modulated, ±25 kHz deviation maximum.
- A *frame* is made up of two *fields* of interlaced horizontal scanning lines. A frame has 525 lines, a field, 262.5 lines.
- *Field frequency* (vertical sweep frequency): 60/s nominal, 59.94/s color.
- *Line frequency* (horizontal sweep frequency): 15 750/s nominal, 15 734.26 color
- *Color subcarrier:* 3.579545 MHz.

FM Broadcast standards (USA):

- *Frequency deviation:* ±75 kHz maximum.
- *Highest modulating frequency:* 15 kHz, left + right channels.
- *Stereophonic system:* subcarrier (suppressed), 38 kHz, left–right channels; pilot subcarrier, 19 kHz.

Preemphasis of higher audio frequencies is applied in FM broadcast and TV sound transmission. An *RC* filter with a critical (–3 dB) frequency of 2120 Hz and 6 dB/decade rolloff is used at the receiver for deemphasis. The corresponding *RC* time constant is 75 μs, and the preemphasis is specified by this term.

5.3 ELECTRICAL WIRING PRACTICE

This section is meant to familiarize the technician with some of the more basic and commonly abused electrical-wiring practices. It is not intended to substitute for the supervision of a qualified and licensed electrician. For complete data on this subject, consult the *National Elec-*

tric Code book, published by the National Fire Protection Association, 470 Atlantic Avenue, Boston, MA 02210, and your local building inspector.

A wiring plan for a residence or small commercial building is shown in Fig. 5-4. In larger systems there may be several branch breaker boxes located remotely from the service entrance and connected to it by feeder lines.

Equipment grounding: All new or replacement wiring should include an equipment-grounding wire (commonly called a ground wire) of a capacity to conduct safely any fault current likely to be imposed. This wire may be bare, or have green or green-and-yellow striped insulation. Terminals to be connected to this ground are green-colored or are marked with the word *green*. Equipment grounding can be achieved by attaching (bonding) from the service-entrance neutral to the street side of the city water meter or to a ground rod not less than 8 ft in length.

The function of the equipment-grounding wire is illustrated in Fig. 5-5. Without this third wire, a short from the appliance or instrument case to the live wire places the case at 115 V to earth ground and presents a shock hazard. The grounding wire keeps the case at earth potential and shorts the entire circuit in the event of a hot-wire-to-case short, thus tripping the circuit breaker.

Bonding must be made between the grounding wire and all conduits, metal boxes, and metal equipment cases.

When replacing outlets or wiring in an existing two-wire system, the grounding terminal of the new outlet must be connected to a cold water pipe or other earth-grounded pipe. It must *not* be connected to the existing neutral wire,

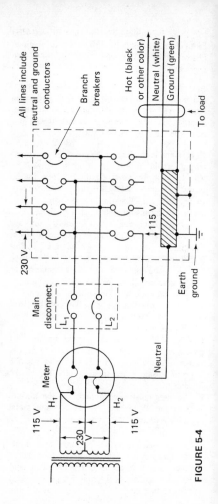

FIGURE 5-4

198

even though this wire is at ground potential. Figure 5-6 shows how this error places the equipment case at 115 V (through the relatively low motor resistance) in the event of an open in the neutral conductor.

Current-carrying conductors: The grounded conductor (commonly called neutral or cold side) should have white or gray insulation. Terminals

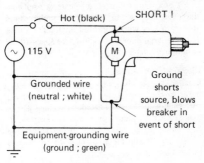

FIGURE 5-5

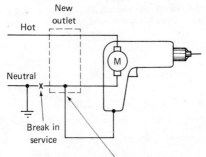

Improper ground makes case hot in event of break in neutral

FIGURE 5-6

to be connected to neutral should be plated with a white or silver metal, or should be marked with the word *white*.

The ungrounded conductor(s) (commonly called the live or hot side) may have any color insulation other than green, gray, or white. Black is commonly used, with red as an alternative. Terminals to be connected to the live wire are generally yellow or copper-colored.

A box must be installed to enclose each conductor splice or tap, except for underground splices made with approved materials. Boxes must be large enough to allow free space for all conductors enclosed.

Electrical wiring circuits: Switches, fuses, and breakers should always be placed in the live side, never in the neutral side of a circuit. Reversing this rule would leave hazardous voltage at the load (to earth ground) even with the switch off, and could place unbalanced loads across the 230-V line, resulting in damage to the lower-power load, as shown in Fig. 5-7.

The shell of a lamp socket is connected to the neutral side (button to hot side) to minimize shock hazard by accidental contact to the

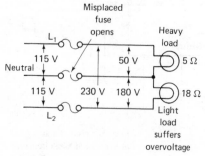

FIGURE 5-7

shell. Figure 5–8 shows the wiring layout for a simple ceiling light and switch. Figure 5–9 shows the layout for two-point control of a single light (the so-called three-way system).

A ground-fault interrupter (GFI) is an inexpensive electronic device that compares the live-wire and neutral-wire currents and trips a breaker if they are unequal. Since most electric shock involves a current path from the live wire to an independent ground (not the neutral wire), the GFI can save a potential shock victim by turning off the power in a matter of milliseconds. Ground-fault protection is required in 115-V outlets installed outside, in a bathroom, in a garage, and on temporary outlets at construction sites. Local codes may also require GFIs in basement and/or kitchen branch circuits.

Load calculation: Dwelling rooms should have outlets placed so that no point on any wall is more than 6 ft horizontally from an outlet. Wall sections 2 or more ft wide should be included in this determination.

At least one outlet should be provided

1. At any kitchen or dining room countertop more than 1 ft wide
2. Adjacent to the bathroom basin
3. In the basement
4. In an attached garage
5. Outdoors

The number of branch circuits required in dwellings, stores, and shops for general lighting and convenience outlets may be determined by allowing 3 watts per square foot of habitable floor space and apportioning the load equally among 15-A branches. For example:

$$1400 \text{ ft}^2 \times 3 \text{ W/ft}^2 = 4200 \text{ W}$$

$$4200 \text{ W} \div 115 \text{ V} = 36.5 \text{ A}$$

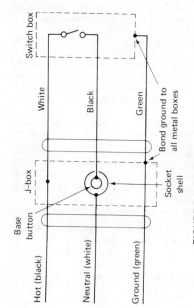

FIGURE 5-8 Simple light switch

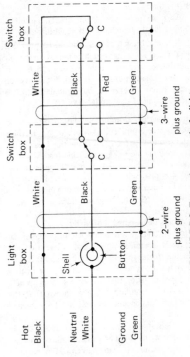

FIGURE 5-9 Two-point control of a light

A *minimum* of three 15-A branch circuits is therefore required for a general lighting load. Local codes may require more.

In addition, a minimum of two 20-A branch circuits are required exclusively for outlets provided for small appliances in the kitchen-pantry–dining area.

Also, at least one 20-A branch circuit is required for receptacle(s) in the laundry area.

Separate branch circuits should be provided for all fixed appliances such as ranges, ovens, air conditioners, dryers, water heaters, space heaters, and heat pumps. Branch-circuit rating should equal or exceed the nameplate ampere or kVA rating of the appliances.

Wiring practice: Conductors should be sized so that the maximum voltage drop to the farthest outlet cannot exceed 3% from the branch circuit breaker, nor 5% from the feeder circuit breakers. Good practice in dwellings is to use No. 14 copper wire for 15-A branch circuits and No. 12 copper wire for 20-A branch circuits.

Where additional outlets and lighting are to be installed, as in building additions or in finishing rooms in basements or attics, it is necessary to install one or more new circuit breakers at the branch-service point and wire the new branches from this point. It is hazardous and generally illegal simply to extend new outlets and lighting from the nearest existing branch, unless load calculations have shown that the maximum branch current will not be exceeded thereby.

Outlets should not be placed over the tub in a bathroom, as this is an invitation to place an appliance where it can fall into the bather's lap.

Wires should be spliced or joined only inside metal boxes, and then they should be joined with wire nuts or solder. Twist-and-tape connections are almost certain to corrode with time, resulting in high resistance and overheating.

Corrosion is especially severe where dissimilar metals, such as copper and aluminum are joined. Crimp- or screw-type connectors and anticorrosion compound should be used on such joints.

When terminating three-wire flexible cords it is a good idea to leave the grounding wire (green) a little longer than the live and neutral wires so that it will be the last to sever if the cord is strained.

5.4 MECHANICAL HARDWARE STANDARDS (see TABLES 5-4 TO 5-6)

TABLE 5-4. Machine-Screw Tap- and Clearance-Drill Sizes

Screw	Tap Drill	Clearance Drill
2–56	50	42
2–64	50	42
3–48	47	36
3–56	45	36
4–40	43	31
4–48	42	31
5–40	38	29
5–44	37	29
6–32	36	25
6–40	33	25
8–32	29	16
8–36	29	16
10–24	25	13/64
10–32	21	13/64
12–24	16	7/32
12–28	14	7/32

**TABLE 5-5. Decimal Equivalents
of Standard Drill Sizes**

Drill Size	Decimal Inches	Drill Size	Decimal Inches	Drill Size	Decimal Inches
70	0.0280	44	0.0860	19	0.1660
69	0.0292	43	0.0890	18	0.1695
68	0.0310	42	0.0935	11/64	0.1709
1/32	0.0313	3/32	0.0938	17	0.1730
67	0.0320	41	0.0960	16	0.1770
66	0.0330	40	0.0980	15	0.1800
65	0.0350	39	0.0995	14	0.1820
64	0.0360	38	0.1015	13	0.1850
63	0.0370	37	0.1040	3/61	0.1875
62	0.0380	36	0.1065	12	0.1890
61	0.0390	7/64	0.1094	11	0.1910
60	0.0400	35	0.1100	10	0.1935
59	0.0410	34	0.1110	9	0.1960
58	0.0420	33	0.1130	8	0.1990
57	0.0430	32	0.1160	7	0.2010
56	0.0465	31	0.1200	13/64	0.2031
3/64	0.0469	1/8	0.1250	6	0.2040
55	0.0520	30	0.1285	5	0.2055
54	0.0550	29	0.1360	4	0.2090
53	0.0595	28	0.1405	3	0.2130
1/16	0.0625	9/64	0.1406	7/32	0.2188
52	0.0635	27	0.1440	2	0.2210
51	0.0670	26	0.1470	1	0.2280
50	0.0700	25	0.1495	A	0.2340
49	0.0730	24	0.1520	15/64	0.2344
48	0.0760	23	0.1540	B	0.2380
5/64	0.0781	5/32	0.1562	C	0.2420
47	0.0785	22	0.1570	D	0.2460
46	0.0810	21	0.1590	1/4, E	0.2500
45	0.0820	20	0.1610		

TABLE 5-6. Sheet-Metal Gages
(Decimal Inches)

Gage No.	American or B & S Aluminum, Copper, Brass	U. S. Standard Iron, Steel, Nickel	Birmingham or Stubs Seamless Tubes; Also by Some Manufacturers for Copper and Brass
10	0.1019	0.140625	0.134
11	0.09074	0.125	0.120
12	0.08081	0.109375	0.109
13	0.07196	0.09375	0.095
14	0.06408	0.078125	0.083
15	0.05707	0.0703125	0.072
16	0.05082	0.0625	0.065
17	0.04526	0.05625	0.058
18	0.04030	0.05	0.049
19	0.03589	0.04375	0.042
20	0.03196	0.0375	0.035
21	0.02846	0.034375	0.032
22	0.02535	0.03125	0.028
23	0.02257	0.028125	0.025
24	0.02010	0.025	0.022
25	0.01790	0.021875	0.020
26	0.01594	0.01875	0.018
27	0.01420	0.0171875	0.016
28	0.01264	0.015625	0.014

5.5 DIGITAL AND COMMUNICATIONS CODES

Binary-coded decimal and 4-bit codes

8-4-2-1 code: In most common use. BCD if 0–9; hex if 0–F.

0	0000	A	1010
1	0001	B	1011
2	0010	C	1100
3	0011	D	1100
4	0100	E	1110
5	0101	F	1111
6	0110		
7	0111		
8	1000		
9	1001		

Excess-3 code: 1's complement of binary gives 9's complement of decimal.

0	0011
1	0100
2	0101
3	0110
4	0111
5	1000
6	1001
7	1010
8	1011
9	1100

Gray code: Only one bit changes between counts.

0	0000	8	1100
1	0001	9	1101
2	0011	10	1111
3	0010	11	1110
4	0110	12	1010
5	0111	13	1011
6	0101	14	1001
7	0100	15	1000

Twos complements: $A + 2s$ complement of $B = A - B$.

0	0000	8	1000
1	1111	9	0111
2	1110	10	0110
3	1101	11	0101
4	1100	12	0100
5	1011	13	0011
6	1010	14	0010
7	1001	15	0001

Alphameric codes (see Tables 5-7 to 5-9)

TABLE 5-7. Baudot Five-Level Teletype

Figure	Letter	54321
	A	00011
?	B	11001
:	C	01110
$	D	01001
3	E	00001
!	F	01101
&	G	11010
#	H	10100
8	I	00110
'	J	01011
(	K	01111
)	L	10010
.	M	11100
,	N	01100
9	O	11000
0	P	10110
1	Q	10111
4	R	01010
Bell	S	00101
5	T	10000
7	U	00111
;	V	11110
2	W	10011
/	X	11101
6	Y	10101
"	Z	10001
CAR RETRN		01000
LINE FEED		00010
LTRS		11111
FIGS		11011
SPACE		00100

Note: 1 = punch hole, 0 = no hole; sprocket-feed holes in tape between columns 2 and 3.

TABLE 5-8. ASCII Seven-Level Code

0	30	A	41	K	4B	U	55	$	24	.	2E
1	31	B	42	L	4C	V	56	%	25	/	2F
2	32	C	43	M	4D	W	57	&	26	:	3A
3	33	D	44	N	4E	X	58	'	27	;	3B
4	34	E	45	O	4F	Y	59	(	28	<	3C
5	35	F	46	P	50	Z	5A	)	29	=	3D
6	36	G	47	Q	51	b	20	*	2A	>	3E
7	37	H	48	R	52	!	21	+	2B	?	3F
8	38	I	49	S	53	"	22	,	2C	LF	0A
9	39	J	4A	T	54	L	23	-	2D	CR	0D

Note: Convert to binary by hex code of preceding section.
b = blank or space

TABLE 5-9. EBCDIC Extended Binary-Coded Decimal Interchange Code

0	F0	c	83	o	96	A	C1	M	D4	Y	E8	)	5D	=	7E
1	F1	d	84	p	97	B	C2	N	D5	Z	E9	:	5E	"	7F
2	F2	e	85	q	98	C	C3	O	D6	¢	4A	-	60	b	40
3	F3	f	86	r	99	D	C4	P	D7	.	4B	/	61	LF	25
4	F4	g	87	s	A2	E	C5	Q	D8	<	4C	'	6B		
5	F5	h	88	t	A3	F	C7	R	D9	(	4D	%	6C		

TABLE 5-9 (Cont.)

6	F6	i	89	u	A4	G	C7	S	E2	+	4E	-	6D
7	F7	j	91	v	A5	H	C8	T	E3	!	4F	>	6E
8	F8	k	92	w	A6	I	C9	U	E4	&	50	?	6F
9	F9	l	93	x	A7	J	D1	V	E5	!	5A	#	7B
a	81	m	94	y	A8	K	D2	W	E6	$	5B	@	7C
b	82	n	95	z	A9	L	D3	X	E7	*	5C	,	70

Zone punch \ Row punch	None	0	1	2	3	4	5	6	7	8	9	3 and 8	4 and 8
None		0	1	2	3	4	5	6	7	8	9	=	.
Zone 12	+		A	B	C	D	E	F	G	H	I	.	)
Zone 11	−		J	K	L	M	N	O	P	Q	R	$	*
Zone 0			/	S	T	U	V	W	X	Y	Z	,	(

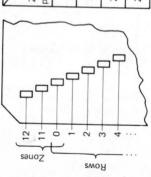

FIGURE 5-10 Computer punched-card code

Hollerith computer punched-card code (see Fig. 5-10)

International Morse (Continental) code (see Fig. 5-11)

A	·—	1	·————
B	—···	2	··———
C	—·—·	3	···——
D	—··	4	····—
E	·	5	·····
F	··—·	6	—····
G	——·	7	——···
H	····	8	———··
I	··	9	————·
J	·———	0	—————
K	—·—	.	·—·—·—
L	·—··	,	——··——
M	——	?	··——··
N	—·		
O	———		
P	·——·		
Q	——·—		
R	·—·		
S	···		
T	—		
U	··—		
V	···—		
W	·——		
X	—··—		
Y	—·——		
Z	——··		

FIGURE 5-11

Police-Radio "Ten" Code (see Table 5-10)

TABLE 5-10. Police 10 Code

Code	Meaning
10-1	Unable to copy—change location
10-2	Signals good
10-3	Stop transmitting
10-4	Acknowledgment
10-5	Relay
10-6	Busy—stand by unless urgent
10-7	Out of service (give location and or telephone number)
10-8	In service
10-9	Repeat
10-10	Fight in progress
10-11	Dog case
10-12	Stand by
10-13	Weather and road report
10-14	Report of prowler
10-15	Civil disturbance
10-16	Domestic trouble
10-17	Meet complainant
10-18	Complete assignment quickly
10-19	Return to . . .
10-20	Location
10-21	Call . . . by telephone
10-22	Disregard
10-23	Arrived at scene
10-24	Assignment completed
10-25	Report in person to . . .
10-26	Detaining subject, expedite
10-27	Driver's-license information
10-28	Vehicle-registration information
10-29	Check records for wanted
10-30	Illegal use of radio
10-031	Crime in progress
10-32	Man with gun
10-33	Emergency
10-34	Riot
10-35	Major crime alert
10-36	Correct time

TABLE 5-10. Police 10 Code (Cont.)

Code	Meaning
10-37	Investigate suspicious vehicle
10-38	Stopping suspicious vehicle (give station complete description before stopping)
10-39	Urgent—use light and siren
10-40	Silent run—no light or siren
10-41	Beginning tour of duty
10-42	Ending tour of duty
10-43	Information
10-44	Request permission to leave patrol for . . .
10-45	Animal carcass in . . . lane at . . .
10-46	Assist motorist
10-47	Emergency road repairs needed
10-48	Traffic standard needs repairs
10-49	Traffic light out
10-50	Accident
10-51	Wrecker needed
10-52	Ambulance needed
10-53	Road blocked
10-54	Livestock on highway
10-55	Intoxicated driver
10-56	Intoxicated pedestrian
10-57	Hit and run
10-58	Direct traffic
10-59	Convoy or escort
10-60	Squad in vicinity
10-61	Personnel in area
10-62	Reply to message
10-63	Prepare to make written copy
10-64	Message for local delivery
10-65	Net message assignment
10-66	Message cancellation
10-67	Clear to read net message
10-68	Dispatch information
10-69	Message received
10-70	Fire alarm

TABLE 5-10. Police 10 Code (Cont.)

Code	Meaning
10-71	Advise nature of fire (size, type, and contents of building)
10-72	Report progress on fire
10-73	Smoke report
10-74	Negative
10-75	In contact with
10-76	En route
10-77	ETA (estimated time of arrival)
10-78	Need assistance
10-79	Notify coroner
10-82	Reserve lodging
10-84	Are you going to meet . . . ; if so, advise ETA
10-85	Will be late
10-87	Pick up checks for distribution
10-88	Advise telephone number to contact
10-90	Bank alarm
10-91	Unnecessary use of radio
10-93	Blockade
10-94	Drag racing
10-96	Mental subject
10-98	Prison or jail break
10-99	Records indicated wanted or stolen

5.6 STANDARD SCHEMATIC SYMBOLS (see Figure 5-12)

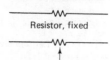

Resistor, fixed

Resistor with adjustable contact; potentiometer

Resistor, 2-terminal adjustable; rheostat

FIGURE 5-12

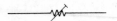

Resistor with preset adjustment; trimmer

Temperature-dependent resistor; thermistor

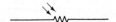

Resistance sensitive to nonionizing radiation
(light, infrared, radio)

Resistance sensitive to ionizing radiation
(x-ray, gamma ray; alpha, beta particle; etc.)

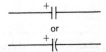

Capacitor: plate length from 3 to 5 times plate
spacing; curved line represents outside plate or
low-potential plate; plus sign used only on
polarized capacitors

Capacitor, with preset
adjust; trimmer

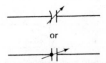

Capacitor, variable. Left plate movable
FIGURE 5-12 (Cont.)

219

Capacitors, mechanically ganged, variable

Capacitor variable differential; one side increases as other side decreases

Capacitor, split stator; both side increase together

Inductor; winding, coils; general and air-core

Inductor, magnetic core

Transformer, general and air-core

Transformer, adjustable coupling

FIGURE 5-12 (Cont.)

Transformer, magnetic-core, nonsaturating

Transformer with electrostatic shield between windings

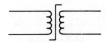

Transformer with saturable core

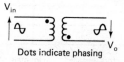

Dots indicate phasing

Wires crossing, not connected

Wires connected

FIGURE 5-12 (Cont.)

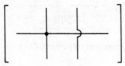

To be avoided

Shielded conductors (2-wire shown)

Coaxial cable

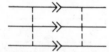

Connectors, engaged. 3-wire male plug (left)
and female receptacle (right) shown

Connector, male plug, 2-wire, nonpolarized

Connector, female receptacle, 3-wire

FIGURE 5-12 (Cont.)

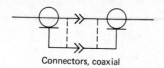

Connectors, coaxial

Common return point

Chassis "ground" connection or common return connection in instruments without conductive chassis

Earth ground or metal vehicle frame

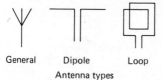

| General | Dipole | Loop |

Antenna types

Antenna counterpoise

FIGURE 5-12 (Cont.)

223

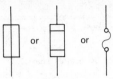

Fuse, general

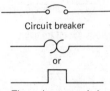

Circuit breaker

Thermal cutout switch

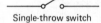

Single-throw switch

Double-throw switch

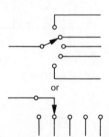

or

Multiposition switch

FIGURE 5-12 (Cont.)

Pushbutton switch, normally open

Pushbutton switch, normally closed

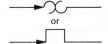

Flasher or self-interrupting switch

Limit switch, normally open (closed by moving machinery)

Limit switch, normally closed

Switch, normally closed, opens after time delay

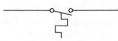

Switch, normally open, closes on rising temperature

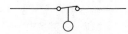

Switch, normally closed, opens on rising liquid level

FIGURE 5-12 (Cont.)

225

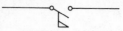

Switch, normally open, closes on increased fluid flow rate

Switch, normally closed, opens on increased pressure

Centrifugal switch, opens on increased angular speed

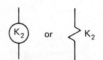

Relay or solenoid coil

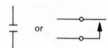

Normally open relay contact (if left symbol is used, make length no more than $1\frac{1}{4}$ times spacing to avoid confusion with capacitor symbol)

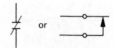

Normally closed relay contacts

FIGURE 5-12 (Cont.)

226

Battery, single-cell left, multicell right; polarity sign optional

AC source, general; oscillator

Thermocouple

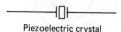

Piezoelectric crystal

Magnetic head, erase (X), record (→), and read (←) functions

Microphone

Loudspeaker

FIGURE 5-12 (Cont.)

Lamp, incandescent, pilot or indicating

Lamp, incandescent, illuminating

Lamp, fluorescent, with heaters

Cold-cathode glow lamp, neon lamp

Meter, general (label inside defines type)

Rotating machine (label defines type)

Semiconductor diode

FIGURE 5-12 (Cont.)

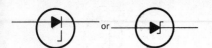

Regulator, avalanche, or zener diode

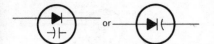

Varactor diode, varicap

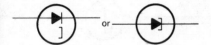

Tunnel (Esaki) diode

Diac, trigger diode

Transistor, NPN

FIGURE 5-12 (Cont.)

Transistor, PNP

Unijunction transistor

FET, junction, N-channel (reverse arrow for P-channel)

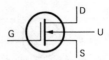

Insulated-gate (MOS) FET, N-channel, depletion type

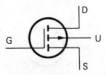

Insulated gate FET, P-channel, enhancement

FIGURE 5-12 (Cont.)

Silicon controlled rectifier (SCR)

Gate turn-off SCR

Triac

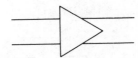

Amplifier, general. Differential inputs and outputs shown

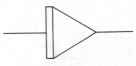

Integrator, general

FIGURE 5-12 (Cont.)

Analog multiplier, divider

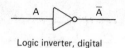

Logic inverter, digital

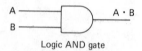

Logic AND gate

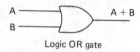

Logic OR gate

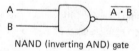

NAND (inverting AND) gate

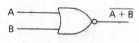

NOR (inverting OR) gate

FIGURE 5-12 (Cont.)

6

Test Procedures

6.1 COMPONENT TESTS

Low-value resistance such as a printed-circuit run or coil winding: Pass a relatively large, measured current through the component from a supply and limiting resistor. Meanwhile, measure the voltage drop across the component and compute $R = V/I$. Clip the voltmeter leads directly to the component to avoid measuring the IR drop in the current-carrying leads [see Fig. 6-1 (a)].

Sensitive resistance, such as a meter coil: Pass a small measured current through a supply and high-value limiting resistance. Measure V across the component and compute $R = V/I$ [see Fig. 6-1 (b)]. Note that using an ohmmeter on microammeter coils may destroy the meter.

High-value resistance: Use a DVM or VTVM with a known high value of input resistance on

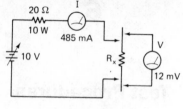

$$R_x = \frac{V}{I} = \frac{12 \text{ mV}}{485 \text{ mA}} = 25 \text{ m}\Omega$$

(a) Low resistance

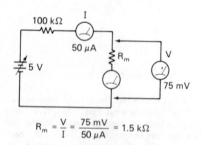

$$R_m = \frac{V}{I} = \frac{75 \text{ mV}}{50 \text{ }\mu A} = 1.5 \text{ k}\Omega$$

(b) Delicate resistance

FIGURE 6-1

the voltage scales to form a voltage divider with the unknown resistance, as in Fig. 6-2(a). The $0.1\text{-}\mu F$ capacitor may be required to keep the high-impedance voltmeter input from picking up too much noise.

Batteries should be tested under load, as shown in Fig. 6-2(b). Where the rated or required output current is known, the value of R_L may be calculated from V_S/I. The formula given with the figure is a rule of thumb based on battery size (volume). Don't forget to calculate the power dissipation requirement for R_L.

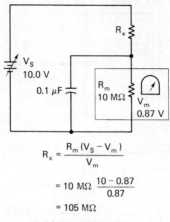

$$R_x = \frac{R_m (V_S - V_m)}{V_m}$$

$$= 10 \text{ M}\Omega \ \frac{10 - 0.87}{0.87}$$

$$= 105 \text{ M}\Omega$$

(a) High resistance

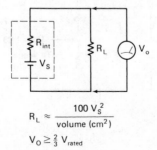

$$R_L \approx \frac{100 \, V_S^2}{\text{volume (cm}^2)}$$

$$V_O \geq \tfrac{2}{3} \, V_{rated}$$

(b) Battery test

FIGURE 6-2

Capacitors from 10 μF to 100 pF can be measured to an accuracy of about 3% with the test setup of Fig. 6-3(a). The generator output is pre-

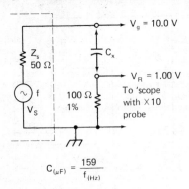

$$C_{(\mu F)} = \frac{159}{f_{(Hz)}}$$

(a) Capacitor measurement

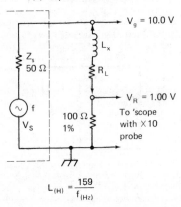

$$L_{(H)} = \frac{159}{f_{(Hz)}}$$

(b) Inductor measurement

FIGURE 6-3

set to 10.0 V on the oscilloscope, and the frequency is varied to produce 1.00 V across the 100-Ω 1% resistor. V_g is then rechecked and adjusted if necessary, and C is calculated from the formula given.

Inductor values from 10 H to 100 μH can be measured using the test circuit of Fig. 6-3(b) and the procedure described above. The range can be lowered to 10 μH if R is changed to 10Ω, but V_g must now be rechecked more carefully. If the Q of L_x is less than 5, accuracy will begin to be seriously impaired.

Inductor Q from 1 to about 200 can be measured with the circuit of Fig. 6-4(a). The inductor value L_x is first measured as in Fig. 6-3(b). Q varies with frequency, so the test frequency f must be selected. Now calculate a capacitance C that will resonate L_x at frequency f. Choose an available high-Q capacitor C_s near to the calculated value and vary f until V_E peaks at f_r. Finally, measure V_g. The ratio V_C/V_g equals Q, since X_L and X_C cancel at f_r and V_g equals V_{RL}. The 2.2-Ω resistor prevents V_g from soaring and dipping wildly around f_r. If Q is very high and the generator waveform contains harmonics, V_g may become very distorted, invalidating the measurement. It may be possible to obtain a reading at a higher frequency (lower C) in this case.

Inductor saturation can be measured with the setup of Fig. 6-4(b). V_S is increased until the oscilloscope, in the ac-coupled mode, shows a marked decrease in 200-Hz signal across the inductor. The formulas given convert I_{sat} to $V_{sat(pk)}$ under sine-wave excitation at frequency f, and to volt-microsecond constant for pulse transformers.

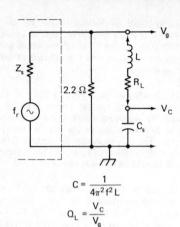

$$C = \frac{1}{4\pi^2 f^2 L}$$

$$Q_L = \frac{V_C}{V_g}$$

(a) Inductor Q

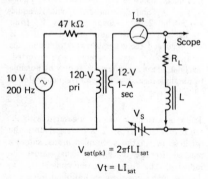

$$V_{sat(pk)} = 2\pi f L I_{sat}$$

$$Vt = L I_{sat}$$

(b) Inductor saturation current

FIGURE 6-4

238

Transformer coefficient of coupling k and turns ratio n can be determined by the measurements and formulas shown in Fig. 6-5. V_1 and V_2 are measured with the generator connected to one winding, and then V_3 and V_4 are measured with the generator across the other winding. If k is near unity, the turns ratio is simply V_2/V_1.

Transformer Thévenin-equivalent winding resistance r_w can be found from ohmmeter tests if skin effect does not come into play [Fig. 6-6 (a)] or from voltage-drop measurements [Fig. 6-6(b)]. R_L is chosen to cause at least a 10% drop from V_{NL} to V_{FL}, and must have an adequate power rating.

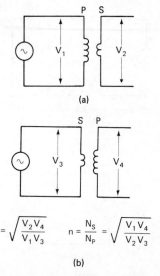

$$k = \sqrt{\frac{V_2 V_4}{V_1 V_3}} \qquad n = \frac{N_S}{N_P} = \sqrt{\frac{V_1 V_4}{V_2 V_3}}$$

(b)

FIGURE 6-5 Turns ratio and coefficient of coupling

239

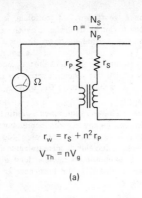

$$n = \frac{N_S}{N_P}$$

$$r_w = r_S + n^2 r_P$$

$$V_{Th} = n V_g$$

(a)

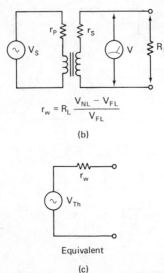

$$r_w = R_L \frac{V_{NL} - V_{FL}}{V_{FL}}$$

(b)

Equivalent

(c)

FIGURE 6-6 Transformer winding resistance

6.2 SEMICONDUCTOR TESTS

Diodes may be tested by simply checking for a low forward and high reverse resistance on a VOM or DVM. A few checks on known-good diodes will permit you to mark your VOM leads for determination of anode and cathode on unknown types. Experience will also allow you to distinguish among silicon, germanium, and Schottky diodes by the progressively lower on-state resistance. Try to settle on one scale of your ohmmeter ($1k\Omega$ or $\times 1$ $k\Omega$) for these tests to avoid confusion. Diodes may be matched for identical on-resistances by this technique.

Diode recovery time t_{rr} for power diodes may be compared with the simple test setup of Fig. 6-7. Fast diodes are required for inverters, switching regulators, and flyback power supplies.

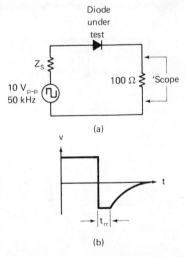

FIGURE 6-7 Diode storage time

Zener or avalanche diode voltage may be checked with the test setup of Fig. 6-8(a). The source voltage should be from 1.5 to 3 times V_z.

Peak inverse or peak-reverse voltage (PIV or PRV) can be measured with the setup of Fig. 6-8(b) if a variable high-voltage supply is available. V_S is increased until the microammeter just begins to indicate a current. V_S then approximates PIV. Be careful not to allow a current of more than a few μA or the diode may be damaged.

Bipolar transistors may be tested as two diode junctions (base–emitter and base–collector). The collector-emitter should show high resistance both ways. Alternatively, the beta of the transistor may be observed with the test shown in Fig. 6-9(a). The ohmmeter provides a voltage source and a current meter for the collector-

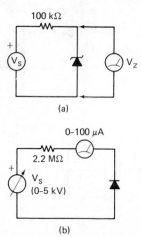

(a)

(b)

FIGURE 6-8 Zener and breakdown voltages

242

emitter. With the base left open, the collector-emitter will conduct only a slight leakage current (high resistance indication). When current is fed from the collector to the base (through a 100-kΩ resistor or through the finger

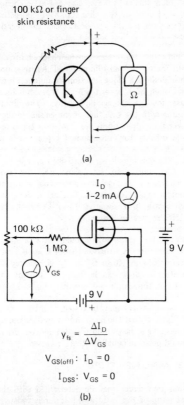

100 kΩ or finger
skin resistance

(a)

$$y_{fs} = \frac{\Delta I_D}{\Delta V_{GS}}$$

$V_{GS(off)}: I_D = 0$

$I_{DSS}: V_{GS} = 0$

(b)

FIGURE 6-9 Transistor and FET tests

skin resistance grasping both leads at once) the collector-emitter will turn on heavily, showing a resistance much less than 100 kΩ.

Power transistors may test more successfully on a lower range of the ohmmeter because of their higher leakage. Use a 1-kΩ collector-base resistor in this case. For PNP transistors, simply reverse the probes to place the negative test voltage at the collector.

Field-effect transistors can be checked for certain faults with an ohmmeter. The gate–source should show a diode junction for junction types and open both ways for MOS types, but in no case low resistance both ways. The drain-source with the gate tied to source should show several hundred ohms for depletion types and near infinity for enhancement types, but in no case near zero.

The test circuit for Fig. 6–9(b) may be used to test almost any small-signal FET for transconductance y_{fs}. Enhancement types will show control of I_D for positive V_{GS} while depletion types will have negative V_{GS}. For P-channel FETS, reverse both supplies.

Unijunction transistors should show a diode junction from E to B_2 and from E to B_1, and several kilohms from B_1 to B_2 either way. A functional check can be made by putting the device in the simple test circuit of Fig. 6–10(a).

SCRs and triacs may be tested with the circuit of Fig. 6–10(b). The lamp should go on when S_1 is pushed and stay on until V_S is removed. Triacs should perform equally with V_S reversed.

6.3 AMPLIFIER TESTS

Input impedance may be determined with the test setup of Fig. 6–11(a) if it is purely resistive. R_v is first set to zero, and V_o is noted. R_v is then

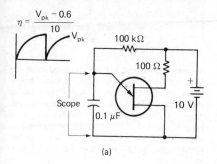

$$\eta = \frac{V_{pk} - 0.6}{10}$$

(a)

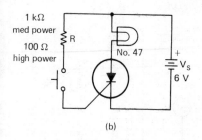

(b)

FIGURE 6-10 UJT and SCR tests

adjusted until V_o drops to one-half of its original value. R_v is then removed and measured. Since one half of V_s is across R_v and the other is across Z_{in}, R_v equals Z_{in}. If Z_{in} is not at least 50 times greater than Z_s of the generator, it will be necessary to shunt the generator with a resistance of about $\frac{1}{50} Z_{in}$. V_o is measured rather than V_{in} because it is larger and because it keeps the oscilloscope from shunting a high value Z_{in}.

Output impedance is measured as shown in Fig. 6-11(b). Measure V_o with no load connected; then add a variable load and adjust until V_o

drops to one-half of its original value. Since one half of V_a appears across Z_o and the other across R_v, $Z_o = R_v$. If distortion becomes evident when R_v is added, lower V_{in}.

Harmonic distortion can be measured with the test of Fig. 6–12 provided that it is not below about 1%. First connect the signal generator to the input of the T-notch filter to determine the depth of the notch and the distortion on the generator sine wave. The output should drop to 0.5% or less of V_s. Then connect the amplifier to be tested between V_s and the notch filter,

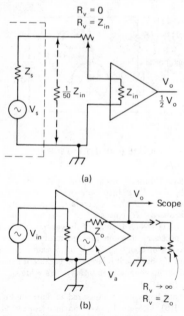

(a)

(b)

FIGURE 6-11 Z_{in} and Z_o tests

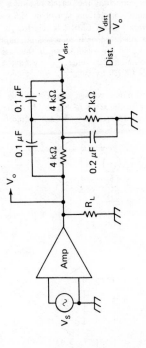

FIGURE 6-12 Harmonic distortion test

$$\text{Dist.} = \frac{V_{dist}}{V_o}$$

and adjust V_s for rated V_o to the load. Measure V_{dist}, preferably with a true rms voltmeter. The notch-filter values shown set the test frequency to about 400 Hz.

Common-mode-rejection ratio CMRR is measured by comparing the output responses for the two input connections shown in Fig. 6-13. If A_v is very high or CMRR is very low, V_1 may become distorted or it may become impossible to see V_2 on the oscilloscope. In this case make V_{in} (common) 100 times V_{in} (differential) and multiply CMRR by 100.

Common-mode voltage limits may be determined with the setup of Fig. 6-14. An ac signal is applied differentially through a transformer,

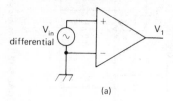

(a)

$$CMRR = \frac{V_2}{V_1}$$

(identical V_{in})

(b)

FIGURE 6-13 CMRR test

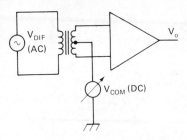

FIGURE 6-14 Common-node voltage limit

and is adjusted to produce an amplifier output not greater than 5% of maximum. A dc voltage is then applied common mode, and increased until the output ac falls below the gain or distortion specifications. The dc supply value is then at the positive common-mode voltage limit. The test is repeated with the dc supply reversed to obtain the negative common-mode limit.

6.4 MISCELLANEOUS TESTS

Hot chassis? A voltmeter test from a suspected chassis to earth ground may be misleading because stray capacitance normally couples relatively harmless currents to a floating chassis. Connect at 47-kΩ $\frac{1}{2}$ -W resistor across the voltmeter and then take the measurement. Leakage currents will produce no more than 10 V across this load.

Percentage modulation (AM): An RF-vs.-time or RF-vs.-AF display can be used to produce one of the patterns of Fig. 6–15, from which percent modulation can be calculated. If the vertical bandwidth of the oscilloscope is less than the RF frequency, the RF will have to be directly coupled to the CRT deflection plates.

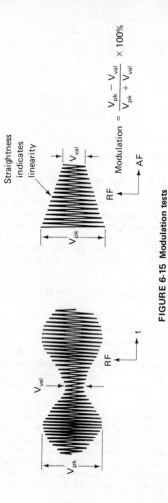

Straightness indicates linearity

$$\text{Modulation} = \frac{V_{pk} - V_{val}}{V_{pk} + V_{val}} \times 100\%$$

FIGURE 6-15 Modulation tests

Glossary of Electronics Terms

ab—prefix for electrical units formed in the cgs system; obsolete.

Acceptor Impurity (semiconductor)—an impurity element containing three electrons in the outer shell, leaving the fourth position vacant to accept a free electron.

Accumulator (computer)—a register that stores the result of an addition, increment, and so on.

Active Component—a component capable of voltage or current gain or switching.

Active Transducer—a transducer producing voltage or current without an external source of electric energy.

A/D Converter—analog-to-digital converter.

Address (computer)—designation of a particular word location in memory or other data register.

AGC—automatic gain control.

ALGOL—ALGOrithmic Language. A compiler language.

Alpha (α)—the ratio of collector to emitter current in a transistor.

Alpha Cutoff (f_α)—the frequency at which alpha drops to 0.707 of its low-frequency value.

Alpha Particle—a particle consisting of two protons and two neutrons.

Alternator—an ac generator.

ALU (computer)—arithmetic logic unit.

Ambient—(adjective) associated with a given environment.

Analog—representation of one quantity by means of another quantity proportional to the first.

Anode—positive electrode.

Architecture (computer)—the conceptual design of computer hardware, including number of accumulators, word length, addressing modes, and so on.

Arithmetic Shift (computer)—shift of digit positions resulting in multiplication or division of the number by a power of the base.

Armature—the moving portion of a magnetic device.

Armature Reaction—distortion of the magnetic field in a motor or generator caused by armature current.

ASCII—American Standard Code for Information Interchange.

Assembler—a computer program that converts user-recognizable coding into machine code on a one-instruction-per-one-instruction basis, also computing address locations and keeping track of data locations by data names.

Audio-Taper Potentiometer—a potentiometer in which resistance changes slowly at counterclockwise rotation and rapidly at clockwise rotation to make response seem linear to the human ear.

Ayrton Shunt—ammeter shunt resistors switched to provide multiple ranges while keeping the same resistance across the meter for optimum damping and not exposing the meter to full circuit current during switch transitions.

Back Diode—a tunnel diode with very low peak tunnel current, used as a low-voltage rectifier conductive in the reverse direction.

Backplane (computer)—conductor interconnections between circuit boards, often by wire wrap.

Ballast Resistor—one in which resistance increases when current increases, tending to keep current constant.

Balun—BALanced to UNbalanced transformer between twin-lead and coaxial lines.

BASIC (Beginner's All-purpose Instruction Code)—a common compiler language similar to FORTRAN.

Baud—signaling speed expressed as level changes per second. 100 pulses per second equals 200 baud.

BCD—binary-coded decimal.

Benchmark Program (computer)—a short program designed to furnish a basis for comparison of two machines.

Beta (β)—ratio of transistor collector current to base current.

BFO—beat-frequency oscillator, used in superhet receivers to heterodyne with code or single-sideband signals to produce audible frequencies.

Bias—voltages or currents required by an element for proper operation. Transistors and tubes require dc bias; transducers, such as recording tape, sometimes use ac bias.

Bidirectional Data Bus—interconnections between two or more registers in which data can be written into or read from the registers via the same set of lines.

Bifet—an integrated circuit having both bipolar and field-effect transistors.

Bifilar—two wires or filaments wound side by side.

arcated Contact—a contact having two fingers for redundant contact.

Binary—having only two possible states, as true/false, 1/0, or $Q/\overline{Q}$ ($\overline{Q}$ = Q-not).

Bipolar Transistor—a transistor using both positive and negative charge carriers in the main current path.

Bit—Binary digIT.

Bit Sliced—a technique of building long-word-length computers by paralleling shorter-word-length processors.

Bootstrap (computer)—a short series of instruction codes that program the machine to read the following codes (a machine with *no* program in memory doesn't even know how to read input data).

Branch Instruction (computer)—an instruction that alters the normal linear sequence of instructions if specified conditions are met.

Bubble Memory—a nonvolatile serial-access integrated-circuit memory using magnetic domains for storage.

Buffer—(1) a circuit that increases the driving capability of a signal; (2) (computer) a memory location for temporary storage of data.

BX Cable—insulated wires in flexible metal armor.

Byte—a binary data unit, generally 8 bits.

Calculator—a device that performs mathematical operations according to instructions and data either entered manually or stored.

Carrier System—a system for sending a number of voice or data channels over a single wire by modulating a number of higher-frequency carrier signals.

Cascade—a number of amplifier stages connected output to input.

Cascode—a common-cathode stage feeding a grounded-grid stage, or a common-emitter stage feeding a common-base stage.

Cathode—negative electrode.

Choke—inductance used to block current above a certain frequency.

Chopper—a device for converting continuous light or current to ac by regular interruption.

Clamper—a circuit used to reestablish a dc level on an ac waveform.

Class-A Amplifier—an amplifier in which plate or collector current flows for the entire 360° of signal swing.

Class-B Amplifier—an amplifier in which plate or collector current flows for essentially one half of the 360° signal swing.

Class-C Amplifier—an amplifier in which plate or collector current flows for considerably less than one half of the 360° signal swing.

Clock—pulse generator supplying synchronizing signals to various components of a digital system.

CMOS (Complementary Metal–Oxide Semiconductor)—a low-power logic family using field-effect transistors.

COBOL (COmmon Business-Oriented Language)—a popular compiler language.

Coherent Radiation—radiation of a single frequency having definite phase relationships between parts of the wave.

Colpitts Oscillator—an oscillator using a two-capacitor voltage divider across the inductor in the tuned circuit.

Common Mode—applied to both sides of a balanced line or amplifier simultaneously.

Compiler—a computer program that translates user-recognizable code into machine code, with one user command often referencing subroutines consisting of many machine instructions.

Computer—a machine that performs mathematical operations according to data and instructions stored in memory, and alters its instructions or branches to new instructions depending upon intermediate results.

Contact Potential—millivoltage appearing across the junction of two dissimilar metals.

Control Unit—that section of a computer which interprets the binary code in the instruction register to activate selected operational functions of other registers or the ALU, such as adding, incrementing, reading, writing, AND-ing, and so on.

Core Storage (computer)—binary memory consisting of tiny saturable magnetic elements.

CPU (computer)—central processing unit.

Cross Assembler (computer)—a program that translates code which runs on one machine into code that will run on another machine.

Curie Point—the temperature above which a magnetic material loses its magnetic properties.

D/A converter—digital-to-analog converter.

Damp—to make an oscillation come to rest.

Darlington Connection—two transistors with the emitter of the first feeding the base of the second, providing a current gain of $\beta_1\beta_2$.

Decoupling—filtering to prevent undesired ac-signal coupling.

Dedicated Computer—a computer programmed for and often wired into a specific task and not intended for other tasks (in contrast to a general-purpose computer).

Degauss—to remove residual permanent magnetization.

Depletion Zone—a portion of a semiconductor near the junction void of charge carriers.

Differential—applied between the two sides of a balanced line or amplifier.

Diffused Junction—a P-N junction formed by diffusion of P-impurity from a vapor into a region of N-type semiconductor, or vice versa.

Digital—a system in which characters or codes are used to represent numbers or physical quantities in discrete steps.

Diplexer—a coupler that allows two transmitters or receivers to operate on a single antenna.

Dipole—an antenna one half wavelength long and split at its center for connection to a feed line.

Direct Addressing (computer)—an instruction for which the address of the data to be operated upon is specified by the word following that instruction.

Direct Coupling—coupling of two stages or circuits by a wire, resistor, or battery.

Disk (computer)—a rotating disk coated with magnetic material and used for bulk storage of data and programs.

DMA (computer)—direct memory access by input/output devices, bypassing the processor.

Documentation—instructions, notes, and diagrams prepared to assist in understanding and use of an instrument or computer program.

Donor Impurity—an impurity that increases the number of free electrons in a semiconductor.

Doping—addition of impurities to a semiconductor.

Double-Base Diode—obsolete term for unijunction transistor.

Double-Diffused Transistor—a transistor in which two P-N junctions are formed by alternately diffusing P- and N-type impurities into an N-type semiconductor.

Double-Precision Arithmetic (computer)—use of two words rather than one to represent each number.

Driver—the amplifier stage preceding the output stage.

Dump (computer)—to transfer the contents of memory into another storage device.

Dynamic RAM—a RAM in which data contents will be lost if not rewritten periodically, typically every 10 ms.

Duplexer—a rapid-switching device that permits use of the same antenna for transmitting and receiving; especially in radar.

Duplex Operation—transmitting and receiving

without noticeable switching between send and receive periods.

Duty Factor—ratio of working time to total time.

Dynamic Braking—slowing an electric motor by connecting it as a generator feeding energy to a resistance or a battery.

Dynamic Loudspeaker—one in which the AF current is carried by a coil that moves with the diaphragm.

Dynamotor—a rotating machine with two windings on one armature for converting one type of power to another, such as 12V dc to 400 V dc.

EAROM (Electrically Alterable Read-Only Memory)—a memory into which data can be written by bit or by word, but much more slowly than data readout. Also called read mostly memory.

ECL (Emitter-Coupled Logic)—a high-speed logic family, also called current-mode logic (CML).

Editor—a program that rearranges digital data, deletes unwanted output such as leading zeros, or inserts standard data such as titles and page numbers.

EDP—electronic data processing.

Effective Radiated Power (ERP)—the product of antenna input power and antenna gain.

Electric—(adjective) using or activated by current in a conductor: as "electric motor."

Electrical—(adjective) related to electric devices but not operated by electric current: as "electrical engineer."

Electricity—1.(noun) a property of electrons and protons, expressed numerically as charge in coulombs; 2.(adjective) electrical: as "electricity book."

Electrolyte—a liquid or paste in which conduction is by flow of ions.

Electronic—(adjective) using or activated by

electric current in semiconductors or evacuated chambers: as "electronic computer."

Electronics—1.(noun) the field of technology that deals with electronic devices; 2.(adjective) related to the field of electronics: as "electronics technician."

Enhancement-Mode FET—an insulated-gate field-effect transistor that is nonconductive at zero gate–source voltage and turns *on* with forward gate bias.

Envelope—the overall shape of an amplitude-modulated waveform, disregarding individual RF variations.

Epitaxial Layer—a thin semiconductor layer condensed from a vapor (grown) onto a thicker substrate layer; used to obtain high purity and regular crystalline structure.

EPROM (Erasable Programmable Read-Only Memory)—a memory into which data can be written with a special programming device, and which is bulk-erasable, usually by exposure to ultraviolet radiation.

Executive Program—a program that manages the loading, running, and outputting of other programs.

Extrinsic Semiconductor — a semiconductor whose electrical properties are determined by the added impurities.

Faraday Shield—an electrostatic shield that passes magnetic and/or electromagnetic fields.

Ferrite—a powdered magnetic material compressed and bonded into a desired shape; used as a core for inductors; capable of high Q at high frequencies.

Fetch (computer)—to retrieve a piece of data from memory.

FIFO (First In, First Out)—a system of storing and retrieving data in a stack.

File (computer)—a block of related data.

Firmwave—1.(microcomputers) programs stored in ROM; 2.(mainframe computers) operating programs supplied with a machine, such as compilers, assemblers, monitors, editors, and so on.

Flag—a single bit of data set or cleared to indicate the presence or absence of a particular condition.

Floating-Point Arithmetic—calculation in which the position of the decimal point is variable.

Flow Chart (computer)—a graphical representation of a computer program, permitting an overview of the program logic.

Flow Solder—mass soldering of printed-circuit boards by moving them over a wave of molten solder.

Fluorescence—emission of light by a substance when exposed to radiation or impact of particles; ceases within a few nanoseconds after bombardment.

Flutter—distortion of sound caused by speed variation of tape or disk during recording or playback.

Fuse—a protective device that opens a circuit on overcurrent.

Fuze—a device used to detonate an explosive charge.

FORTRAN (FORmula TRANslater; computer)—a popular compiler program designed primarily for science and mathematics.

Gate-Turnoff Switch—a switching device similar to an SCR but able to be turned off from its gate terminal.

Gaussian Distribution—a continuous symmetrical distribution of data about the mean; normal distribution; bell curve.

Glitch—a noise spike.

Growler—a test device used to identify shorts in the armature of a motor or generator.

Gunn Diode—a diode that produces gigahertz oscillations when dc-biased at the proper voltage.

Half Adder—a logic circuit that adds two binary digits producing sum and carry outputs, but is unable to handle a carry input. Two half adders can be connected to form a full adder which will handle carry input.

Hall Effect—development of a voltage across a metal or semiconductor block placed in a magnetic field.

Handshaking—data transfer in which the receiving register sends a *ready* and/or *acknowledge receipt* signal back to the sending register via a separate wire.

Hard Copy (computer)—paper or card printout, as opposed to a cathode-ray-tube display.

Hardware—the physical computer and associated machines.

Hard-wired—inherent in the electronic circuit configuration, as opposed to *programmed;* unalterable.

Harmonic—a sinusoid having a frequency that is an integral multiple of the fundamental frequency.

Hartley Oscillator—an oscillator characterized by a tapped coil in the tuned circuit.

Hermetic Seal—a seal preventing the passage of air, water vapor, or other gases.

Heterodyne—to mix two ac signals of frequencies f_1 and f_2 in a nonlinear device, producing additional output frequencies $(f_1 + f_2)$ and $(f_1 - f_2)$.

Hexadecimal—the number system in base 16, counted 0, 1, 2, 3, 4, 5, 6, 7, 8, 9, A, B, C, D, E, F. Often used in computing, where two "hex" digits represent one byte of data.

Hole—a vacancy in the electron structure of a semiconductor; regarded as a mobile positive charge carrier.

Holography—three-dimensional photography using laser light. The effect of viewing a hologram is as if one were looking through a window at a three-dimensional object.

Hybrid Integrated Circuit—a combination of integrated circuits with small, often unencapsulated discrete components.

Hysteresis—a lag effect similar to mechanical friction, sometimes expressed by the graphic words "slop," "stickiness," or "dead zone."

Ignitron—a high-power controlled rectifier using a pool of mercury to emit ions for conduction.

$I^2 L$ (Integrated Injection Logic)—a direct-coupled bipolar-transistor logic family presently used primarily in large-scale specialized integrated circuits.

Image—a spurious signal encountered in super-heterodyne receivers (at a frequency $f_r + 2f_{if}$ when the local oscillator is above the received frequency f_r).

Immediate Addressing—an addressing mode in which the operand datum itself appears in the program at a location immediately following the instruction.

Indexed Addressing—an addressing mode in which the address of the operand data is obtained by adding the *base address* (given immediately after the instruction) to the index register (which may be incremented or decremented by another instruction). Used to step through a file of data on successive passes through a program loop.

Indirect Addressing—an addressing mode in which a *pointer address* immediately follows the instruction. The operand address is contained in the address pointed to. Used to permit modifying the operand address without modifying the program.

Input/Output (I/O)—related to the problem of getting data into and out of a computer.

Instruction—a set of binary digits interpreted by the computer to activate specific machine functions, such as add, shift, store, increment, etc.

Instruction Cycle—the set of machine cycles required to execute a given instruction. Typically, one to four machine cycles.

Instrument Transformer—one that transfers voltage, current, or phase information from a high-level primary to a low-level secondary for purpose of measurement.

Integrated—a multitude of parts brought together and made one.

Interface—the circuitry or connections between a computer and an I/O device such as a thermistor or a line printer.

Intermediate Frequency—the difference frequency $f_1 - f_2$ produced in a superheterodyne receiver. Standard IFs are 455 kHz for AM radio, 10.7 MHz for FM broadcast, and 45.75 MHz for TV picture carrier.

Intermodulation—distortion produced when two signals of different frequencies appear in an amplifier that is not perfectly linear. As a form of heterodyning, it produces sum and difference frequencies.

Interpreter—a program that executes compiler-language statements immediately, one by one, instead of assembling all statements before execution.

Interrupt—the process of halting execution of a main program, storing intermediate results and register conditions, executing a short priority program, restoring the conditions prevailing before the interrupt, and resuming execution of the main program.

Intrinsic Semiconductor—essentially pure semiconductor whose properties are not determined by impurities.

Inverter—a device for converting dc to ac by switching dc alternately in inverted polarity.

Ion—an atom or molecule that has acquired a charge by gaining or losing one or more electrons.

Isolation Transformer—a one-to-one transformer used to isolate equipment at the secondary from earth ground which constitutes one side of the ac main at the primary.

Isotopes—atoms having the same number of protons but a different number of neutrons.

Isotropic—having identical properties in all directions.

Iterative Process—a mathematical technique for calculating a desired result by successively closer approximations.

Jack—a connector into which a plug may be inserted.

Klystron—a UHF electron tube in which an electron beam is bunched by electric fields and fed into a resonant cavity.

Lagging—occurring later in time.

Laplace Transform—a technique for solving circuit equations by reducing differential equations to algebraic equations.

Laser (Light Amplification by Stimulated Emission of Radiation)—an electron device that produces a beam of coherent light.

Latch—a storage element that retains digital data while the data source is in a state of change.

Lattice—a pattern of positions on a regular grid of lines.

Lead Time—the time required for a manufacturer to develop a product and tool up to be ready to produce it at a given time.

LeClanche Cell—the common carbon-zinc dry cell.

Light Pipe—a transparent plastic rod that transmits light from one end to the other, whether rigid or flexible, straight or bent.

Lissajous Figure—the pattern on an oscilloscope when two sine waves of related frequencies are applied to the vertical- and horizontal-deflection systems.

Load—the device that receives the output of a signal source.

Logic Function—an expression of the relationship between binary inputs and output of a circuit, including AND, OR, NOT, NAND, NOR, and XOR (exclusive or).

Long Tail—an emitter follower using an emitter-current source or a high-value emitter resistance to improve linearity.

Look-Ahead Carry—use of separate logic to generate the carry output in adders and counters; eliminates waiting for the count to ripple through all bits and permits fast cascading of units.

Lookup Table—a computer technique in which a required function (say, temperature) is obtained from data stored in a file and fetched out in response to an input variable (say, thermistor voltage).

Loop—a series of instructions executed a number of times with different data each time.

Loran—LOng RAnge Navigation using pulse transmissions from widely spaced stations operating between 1.8 and 2.0 MHz.

Machine Cycle (computer)—one complete fetch-execute cycle; may consist of one or several clock cycles.

Machine Language—program and data in a form immediately usable by the computer, usually binary.

Macroinstruction—an instruction given to the computer as a single instruction but which may require several machine cycles executing a number of different operations; accomplished by microprogramming.

Magnetostriction—a change in length of a magnetic rod when placed in a magnetic field; Joule effect.

Magnetron—a UHF power-oscillator electron tube using magnetic fields and cavity resonators.

Mainframe—a full-sized computer, as opposed to a mini or micro.

Majority Carriers—electrons in N-type material; holes in P-type material.

Maser (Microwave Amplification by Stimulated Emission of Radiation)—a low-noise amplifier similar to the laser in principle.

Masking—coating selected areas of a semiconductor wafer, leaving other areas exposed for diffusion, etching, or metallization.

Master-Slave Flip-Flop—a pair of flip-flops treated as a unit; the master changes state on the rising clock pulse and the slave follows on the falling clock pulse. This removes ambiguity where the clock falls during an input transition.

Maximum Usable Frequency—the highest frequency that will be reflected by the ionosphere at a given time.

Memory-Mapped—a computer architecture that addresses I/O devices as if they were memory locations, and provides ready access to and control of data in all memory locations. Characteristic of the 6800 and 6502 microprocessors.

Mesa Transistor—a precursor of the planar transistor in which a wafer is etched down around protruding base and emitter regions.

Microcomputer—a fully operational computer built around a microprocessor, whether consisting of separate chips or integrated on a single chip.

Microprocessor—a central processing unit (ALU and control system) in large-scale integrated-circuit form, requiring only the addition of memory, I/O devices, and sometimes a clock to form a microcomputer.

Microprogramming—a short series of instructions hard-wired into a microprocessor to permit it to perform several fetch-execute cycles in response to one instruction. Permits relatively powerful commands, such as ADD, to be included in a microprocessor instruction set.

Miller Effect—a decrease in input impedance of an amplifier caused by amplified voltage across an impedance from output to input, which in turn produces high input current.

Miller Integrator, Miller Runup—a linear-ramp generator which operates by charging a capacitor connected from output to input of an inverting amplifier.

Minicomputer—a computer midway in size, sophistication, and capability between a micro-computer and a mainframe.

Mixer—(1) a linear device used to combine several audio or video signals in any desired proportion; (2) a nonlinear device used to heterodyne two signals.

Modem (MOdulator-DEModulator)—a device that connects a computer to a telephone line, usually by frequency modulation of an audio tone.

Modulator—a device that varies the amplitude, frequency, or phase of an ac signal.

Monitor (computer)—a housekeeping program that permits the computer to communicate with its I/O devices, step through addresses, modify programs, check register conditions, single-step through programs, and so on.

Monolithic Integrated Circuit—an integrated circuit fabricated from a single chip of semiconductor material, usually silicon.

Motherboard—a board containing a number of printed-circuit sockets and serving as a backplane.

Multiplex (MUX)—any of several techniques for sending multiple communications channels over a single wire pair or radio link.

Narrow-band FM—frequency modulation with deviations less than ±15kHz.

NC—(1) normally closed; (2) no connection.

Negative Resistance—a property of tetrode tubes, tunnel diodes, and some other semiconductors wherein increased voltage causes decreased current over a limited range.

Nesting (computer)—a programming technique of putting one loop inside another loop.

Neutralization—a technique for combating self-oscillation in an amplifier by providing negative feedback to cancel positive feedback through stray capacitance.

Nondestructive Readout—reading memory data without erasing those data.

Normalize—to multiply all data by a factor such that the reference data level is unity.

Null—position of minimum or zero response or output.

Nybble—one-half byte; four binary digits.

Object Program—program in a form usable by the computer; as opposed to source program.

Octal—the number system to base eight, counted 0, 1, 2, 3, 4, 5, 6, 7, 10, 11, 12, ...

OEM—Original Equipment Manufacturer.

Operand (computer)—the data upon which an instruction is to operate.

Operating Point—the combination of voltage and current at which a tube or transistor is biased.

Overlaying (computer)—transferring programs from bulk to high-speed storage during processing of earlier parts of the program to make maximum use of high-speed storage.

Overtone Crystal—a crystal designed to oscillate on a harmonic of its natural fundamental frequency.

Pad—an attenuator used to stabilize impedances on a line.

Page—a block of memory addressable by one machine word: for example, a 256-word block in a microcomputer having an 8-bit word length.

Panoramic Receiver—a receiver having a CRT display of signal strength vs. frequency for all signals near the received frequency.

Parametric Amplifier—a UHF amplifier whose operation is based on change in the reactance of a tube or semiconductor with a locally generated "pumping" voltage.

Parasitic Element—an antenna element that reflects or reradiates energy but is not directly connected to the transmission line.

Parasitic Oscillation—an undesirable high-frequency oscillation caused by stray inductance or capacitance and at a frequency unrelated to the operating frequency.

Parity—a binary digit added to a word to make the sum of all bits always 1 or always 0. Used to catch machine errors.

Passivation Layer—an oxide layer deposited on the surface of a transistor or IC to preserve it from contamination.

Passive Component—a component that is not capable of amplification or switching action.

Peltier Effect—heating or cooling of a junction of dissimilar metals when a current is passed through them.

Peripherals (computer)—hardware used in conjunction with a computer, such as card readers, tape storage devices, displays, and so on.

Phase—the position of one waveform with respect to another of the same frequency, expressed in degrees, with 360° representing one complete cycle.

Phase-Locked Loop—a circuit in which a voltage-controlled oscillator follows the frequency (or multiple thereof) of a reference signal.

Photoresistive—changing resistance with light intensity.

Photovoltaic—generating a voltage as a result of light radiation.

Piezoelectric Effect—generation of a voltage in response to a mechanical vibration, and conversely, mechanical vibration in response to an applied ac voltage.

Pinchoff Voltage—the reverse gate–source voltage, which reduces channel current of an FET to a specified near-zero level.

Pink Noise—noise having more power content toward the low end of the frequency spectrum.

Planar Process—a process of manufacturing transistors and ICs in which junctions are diffused into a flat epitaxial layer.

Point-Contact—an early process for diode and transistor fabrication in which wire points are bonded to a base piece of doped semiconductor.

Potential Barrier—a semiconductor region in which charge carriers are repelled and may be turned back unless the external applied voltage is sufficient.

Potting—a rubber or plastic insulating compound in which an assembly may be encapsulated for protection from vibration, moisture, etc.

Program—a sequence of instructions to be followed by a computer.

Program Counter (computer)—a register that stores the address of the next instruction to be executed.

PROM (Programmable Read-Only Memory)—a memory into which data can be written only once, either at the factory by masking bits to 1 or 0, or in the field by blowing tiny internal fuse links in a programming fixture.

Propagation Time—the time required for a signal to reach one point from another, be it through free space, a transmission line, an amplifier, or a logic gate.

QA—Quality Assurance; also QC (quality control).

Quadrature—a 90° phase relationship.

Quiescent—at rest; bias conditions without signal.

Race—an ambiguous condition when two logic levels are changing states nearly simultaneously, the result depending upon which one changes first.

Radar—RAdio Detection And Ranging.

Radix—the base of a number system.

RAM (Random-Access Memory)—a memory in which each word can be read as quickly as any other word. Commonly used to denote a *read/write* memory as opposed to a read-only memory.

Raster—the pattern of scanning lines covering the CRT face in television.

Read—to pick up data from a register or memory location.

Real Estate—area on a printed circuit or IC chip available for placement of components.

Reflectometer—a directional coupler used to measure reflected waves or standing-wave ratio on a transmission line.

Register—a binary storage device, generally for one word of data.

Register Oriented—a computer architecture in which memory is accessed relatively infrequently and most operations are performed in a number of *working registers*. Characteristic of the 8080 microprocessor.

Rel—reliability.

Relative Addressing (computer)—an instruction mode that references a memory location in terms of distance from the current program counter contents, rather than in absolute terms.

Relaxation Oscillator—an oscillator whose frequency is determined by the charging time of an *RC* circuit.

Relay Rack—a frame to accommodate standard 19-in.-wide panels for equipment mounting.

Repeater—a device that receives weak signals and transmits corresponding stronger signals, sometimes on a different frequency and sometimes after signal processing.

Residual Magnetism—that remaining in an electromagnet after the magnetizing current has been reduced to zero.

Resonance—the condition wherein the frequency of an externally applied force equals the natural oscillation frequency of a system.

Rheostat—a two-terminal variable resistor, especially one of high power rating.

Ringing—a damped oscillation following a step change in input.

ROM (Read-Only Memory)—a memory into which data have been permanently programmed and into which new data cannot be written.

Salient Pole—in a motor or generator, a magnetic pole that protrudes from, rather than blends into, the cylindrical shape of rotation.

Sampling Oscilloscope—an oscilloscope capable

of displaying repetitive signals well above 1 GHz by fast switching of diodes to obtain samples of the signal at progressively delayed points on the wave.

Saturation—the point at which increasing one quantity no longer has an effect on a second quantity; commonly applied to base current vs. collector current and to magnetizing current vs. magnetic flux.

SCA (Subsidiary Communication Authorization)—subscription music service transmitted by subcarrier from FM broadcast stations.

Second Detector—the demodulator that converts the IF signal to AF in a superheterodyne receiver.

Seebeck Effect—voltage generated across a junction of two dissimilar metals as a result of temperature differences between two such junctions.

Serial (computer)—handling data sequentially rather than simultaneously.

Servo System—a feedback control system in which the output response is a mechanical position.

Set Point—the value of a controlled variable to be maintained by a process controller; demand point.

Short Circuit—an undesired or temporary low-resistance path.

Sidebands—a span of frequencies on either side of a modulated carrier produced when modulation distorts the perfect sine-wave shape.

Significant Figure, Significant Digit—a part of a number whose true value is known, and is not simply the result of mathematical computation or zeros serving as placeholders.

Simplex Operation—radio communication requiring manual or automatic switching between talk and listen periods.

Single-Ended—having the signal appearing from one line to ground, rather than differentially between two lines balanced around ground.

Single Sideband—a form of amplitude modulation in which the carrier and one sideband are removed and all transmitter power is concentrated in the other sideband; advantages include: half the bandwidth, freedom from carrier heterodyne (howl), and better immunity from fading. Disadvantages: circuit complexity, reduced fidelity, and critical tuning.

Slewing—moving as rapidly as possible from one point to another.

Snap-Action Diode—a diode that transitions abruptly from conduction to nonconduction.

Software (computer)—programming, especially problem-oriented, as opposed to housekeeping programs.

Solid-State Device—a device that controls electric current within solid materials, as opposed to vacuum, gases, or liquids.

Sonar—SOund Navigation And Ranging.

Source Program—user-readable statements before they have been converted into machine language.

Spurious Signals—unwanted signals of chance or questionable origin.

Square-Law Detector—one whose output is proportional to the square of the input.

Squirrel-Cage Rotor—a motor armature containing copper rods connected to copper end disks, the entire assembly embedded in a relatively low conductivity iron core.

Stack (computer)—an area of memory designated to store intermediate results and operating-register contents during an interrupt.

Stagger Tuning—adjusting a number of tuned circuits to slightly different frequencies to give a wider, flat-topped overall response curve.

Standing-Wave Ratio—the ratio of maximum to minimum voltage or current along a transmission line; equal to unity if load, line, and source impedances are equal and purely resistive.

Static RAM—a RAM in which memory contents are preserved as long as continuous voltage is maintained at the supply pin.

Strain Gage—a resistive element designed to be attached to the surface of a mechanical-support member and to change resistance in a predictable way when that member is subjected to stress.

Strobe—to gate *on* and *off* at a regular rate.

Stub—a short section of transmission line connected in parallel with the main line and used for tuning, trapping, or impedance matching.

Subcarrier—a fixed ultrasonic frequency that modulates an RF carrier (in addition to normal AF modulation) and is itself modulated by a second audio channel.

Subroutine—a section of a program that performs a well-defined and frequently used function. The program counter leaves the main program, counts through the subroutine, and then returns to the main program.

Substrate—the base upon which a transistor or IC is fabricated.

Superheterodyne Receiver; Superhet—a receiver in which all received signals are converted to a fixed intermediate frequency for amplification and selectivity before demodulation.

Surge—a brief increase in voltage or current.

Symbolic Programming—use of user-recognizable alphanumeric codes for instructions and addresses, which can then be converted directly to machine code.

Synchro—a device used to duplicate an angular position at a remote location via multiwire connections.

Synchronous—in time coincidence; in step.

Sync Pulse—in facsimile and television, a pulse transmitted at the end of a line or field to keep the transmitter and receiver in synchronism.

Syntax—the rules of a programming language.

Synthesize—a technique for developing a large number of frequencies from a few master

oscillators using phase-locked loops, heterodyning, and digital frequency division.

System—a combination of several pieces of equipment organized to perform a specific task.

Thermal Runaway—a condition in a transistor
in which heating causes increased collector
current, which causes further heating, in a
spiral ending in saturation or the destruction
of the transistor.

Thermionic—producing emission of electrons by
heating.

Thick-Film Circuit—a circuit in which resistors,
capacitors, and conductors are printed or
painted onto a substrate, often by silk-screen
printing.

Thin-Film Circuit—a circuit in which resistors,
capacitors, conductors, and/or semiconductors
are deposited onto a substrate molecule-by-
molecule in an evacuated chamber.

Thyristor—a family of switching semiconductor
devices, including SCRs, triacs, and diacs.

Toggle—snap action from one state to another.

Transducer—a device that converts energy from
one form to another, especially one that converts some physical quantity to electric current or voltage for purposes of measurement
or control.

Transient—a short pulse or oscillation, as opposed to steady-state conditions.

Transistor (TRANSfer resISTOR)—an active
semiconductor having three electrodes.

Trap—a tuned circuit used to eliminate an undesired frequency.

Traveling-Wave Tube—a UHF electron tube in
which a wave traveling along a helix interacts
with an electron beam traveling down the center of the helix.

Trimmer—a small capacitor or resistor adjustable
by screwdriver or thumbwheel for purposes of
alignment.

Tri-state—logic circuitry having three distinct

output states: high voltage (V_{CC}), low voltage (ground), and high impedance (output floating).

TTL (Transistor-Transistor Logic, T^2L)—A popular family of bipolar digital integrated circuits.

Tunnel Diode; Esaki Diode—a semiconductor diode exhibiting negative resistance between approximately 0.2 and 0.4 V forward bias.

Two's Complement—the result of complementing a binary number (changing all 1's to 0 and all 0's to 1) and adding 1. Adding the two's complement of B to A is equivalent to subtracting B from A.

UART (Universal Asynchronous Receiver Transmitter)—a device that converts one word of parallel data bits (at the computer) to a string of serial data bits transmitted over a single line, and vice versa for receiving.

Varactor—a semiconductor diode whose capacitance decreases with increased reverse bias.

Vector (computer)—the starting address of a routine, especially one to which the computer is directed by an interrupt request.

Vernier—a device used for making fine adjustments or readings of a parameter.

V/F—Voltage-to-Frequency conversion.

Video Frequency—a band extending from a few hertz to about 5 MHz.

Virtual Ground—not actually grounded, but at ground potential for purposes of most calculations.

Volatile Memory—read/write memory whose contents are lost if the power supply is interrupted.

White Noise—random noise, having equal energy at all frequencies.

Wire Wrap—a solderless connection made by winding a bare end of solid wire around a square post having sharp corners.

Word (computer)—a set of binary digits treated as a unit by the computer in data-transfer and arithmetic operations. Typical word lengths are: microcomputer, 8 bits; minicomputer, 16 bits; mainframe computer, 64 bits.

Write—to store data in a register or memory location.

X-Y Plotter—a servo-controlled pen that draws a graph of two variables input as voltages.

Yagi Antenna—a directive antenna consisting of a driven element, a reflector dipole, and several director dipoles.

Zero-Page Addressing—an instruction mode in which the memory page of the operand is assumed to be zero. Frequently accessed data are stored on page zero to shorten the main program by eliminating the need to specify the data page.

Index